John A. Fleming

Electrical Laboratory Notes and Forms

Elementary series - advanced series

John A. Fleming

Electrical Laboratory Notes and Forms
Elementary series - advanced series

ISBN/EAN: 9783337276454

Printed in Europe, USA, Canada, Australia, Japan

Cover: Foto ©berggeist007 / pixelio.de

More available books at **www.hansebooks.com**

Electrical Laboratory Notes and Forms.

ELEMENTARY SERIES.

1. THE EXPLORATION OF MAGNETIC FIELDS.
2. THE MAGNETIC FIELD OF A CIRCULAR CURRENT.
3. THE STANDARDIZATION OF A TANGENT GALVANOMETER BY THE WATER VOLTAMETER.
4. THE MEASUREMENT OF ELECTRICAL RESISTANCE BY THE DIVIDED WIRE BRIDGE.
5. THE CALIBRATION OF THE BALLISTIC GALVANOMETER.
6. THE DETERMINATION OF MAGNETIC FIELD STRENGTH.
7. EXPERIMENTS WITH STANDARD MAGNETIC FIELDS.
8. THE DETERMINATION OF THE INTERPOLAR FIELD OF AN ELECTROMAGNET WITH VARYING LENGTHS OF AIR GAP.
9. THE DETERMINATION OF RESISTANCE AND TEMPERATURE COEFFICIENTS WITH THE POST OFFICE PATTERN OF WHEATSTONE'S BRIDGE.
10. THE DETERMINATION OF ELECTROMOTIVE FORCE BY THE POTENTIOMETER.
11. THE DETERMINATION OF CURRENT STRENGTH BY THE POTENTIOMETER.
12. A COMPLETE TEST OF A PRIMARY BATTERY.
13. THE CALIBRATION OF A VOLTMETER BY THE POTENTIOMETER.
14. A PHOTOMETRIC EXAMINATION OF AN INCANDESCENT LAMP.
15. THE DETERMINATION OF THE ABSORPTIVE POWERS OF SEMI-TRANSPARENT SCREENS.
16. THE DETERMINATION OF THE REFLECTIVE POWERS OF VARIOUS SURFACES.
17. THE DETERMINATION OF THE ELECTRICAL EFFICIENCY OF A SMALL ELECTROMOTOR.
18. THE EFFICIENCY TEST OF A MOTOR.
19. THE EFFICIENCY TEST OF A MOTOR-DYNAMO.
20. TEST OF A GAS ENGINE AND DYNAMO PLANT.

ADVANCED SERIES.

21. THE DETERMINATION OF THE SPECIFIC ELECTRICAL RESISTANCE OF A SAMPLE OF WIRE.
22. THE MEASUREMENT OF LOW RESISTANCES BY THE POTENTIOMETER.
23. THE MEASUREMENT OF ARMATURE RESISTANCES.
24. THE STANDARDIZATION OF AN AMPERE-METER BY COPPER DEPOSIT.
25. THE STANDARDIZATION OF A VOLTMETER BY THE POTENTIOMETER.
26. THE STANDARDIZATION OF AN AMMETER BY THE POTENTIOMETER.
27. THE DETERMINATION OF THE MAGNETIC PERMEABILITY OF A SAMPLE OF IRON.
28. THE STANDARDIZATION OF A HIGH TENSION VOLTMETER.
29. THE EXAMINATION OF AN ALTERNATE CURRENT AMMETER.
30. THE DELINEATION OF ALTERNATE CURRENT CURVES.
31. THE EFFICIENCY TEST OF A TRANSFORMER.
32. THE EFFICIENCY TEST OF AN ALTERNATOR.
33. THE PHOTOMETRIC EXAMINATION OF AN ARC LAMP.
34. THE MEASUREMENT OF INSULATION AND HIGH RESISTANCE.
35. THE COMPLETE EFFICIENCY TEST OF A SECONDARY BATTERY.
36. THE CALIBRATION OF ELECTRIC METERS.
37. THE DELINEATION OF HYSTERESIS CURVES OF IRON.
38. THE EXAMINATION OF A SAMPLE OF IRON FOR HYSTERESIS LOSS.
39. THE DETERMINATION OF THE CAPACITY OF A CONCENTRIC CABLE.
40. THE HOPKINSON TEST OF A PAIR OF DYNAMOS.

Particulars will be found overleaf. The Notes are arranged by Dr. J. A. FLEMING, of University College, London, and are published by "THE ELECTRICIAN" Printing and Publishing Company, Limited, Salisbury Court, Fleet Street, London, England.

INSTRUCTIONS

FOR THE USE OF THE APPENDED

Electrical Laboratory Notes and Forms.

————— •—•—• —————

T HESE Laboratory Notes and Forms are intended for the use and assistance of Students and Demonstrators in Electrical Laboratories. They are not designed to supersede oral or text-book teaching, but to aid the Student in acquiring a habit of immediately and systematically recording the results of observations made in the Laboratory. With the object of obviating a constant repetition of instructions the first two or three pages of each Form are occupied with a brief account of the experiment or measurement to be made, and with practical notes on the precautions necessary to be observed in carrying it out. The Student should read this part carefully before beginning the experiment, and the Teacher may amplify it, as much as necessary, with verbal explanations. It is very desirable to give the Student references to passages in the text-books used, bearing on the particular experiment in progress. It is also extremely advisable that the Student should draw, in every case, a careful diagram showing the arrangement of the apparatus and circuits for each experiment. In all instances, where possible, the results of observations should be set out in curves on sectional paper *at the time* of making the observations, in order that errors may be quickly noticed and remedied. One object of Laboratory teaching being to train the powers of observation in methods of accurately recording the result, a systematic record of all such work should be required from the Student in every case.

In the Bound Volumes, additional blank ruled pages are given between each Paper for further observations, if necessary.

The Sectional Paper used should be of good quality, accurately divided, and of the same size as the "Laboratory Notes and Forms."

Where loose copies of the Notes and Forms are used, it is best to keep them in a Portfolio or similar case.

ELECTRICAL LABORATORY NOTES AND FORMS.

No. 1.—ELEMENTARY.

Name   *Date*

The Exploration of Magnetic Fields.

The apparatus required for these experiments is a circular coil of insulated wire and a magnetic compass box with a degree scale and short magnetic needle having a larger index needle attached to it. The box is arranged to slide along a rod fixed perpendicularly to the plane of the coil, and so that the needle is always on the axis of the coil. The magnetic needle should not be larger than one-tenth the diameter of the coil.

The Student is recommended to sketch the arrangement of the apparatus.

At every point in the neighbourhood of a magnet or conductor carrying a current there is a *magnetic force*, and the region within which this force manifests itself is called a *magnetic field*. At each point in a magnetic field the magnetic force has a certain *direction* and *magnitude*. If a very long magnetic needle is held with one pole at any point in a magnetic field and the other pole removed quite away from that point, the magnetic force tends to urge the pole held in that field in a certain direction, with a force which is proportional to the product of the strength of the pole and the strength of the field at that point. If a very small magnetic needle, pivoted at the centre, is held at any point in a field, two equal and opposite forces act upon its ends. If l is the magnetic length of this needle, m the strength of each pole, and H the strength of the field, then, when the small needle is held so that its length or axis is at right angles to the direction of the field, the *couple* or *torque* acting on it is equal to $m\,l\,$H.

The product $m\,l$ is called the *moment of the magnet*, and is represented by M. Hence, the couple in such a case is numerically equal to the product of the moment of the magnet and the strength of the field $=$ M H.

Hence, we may define a magnetic field of unit strength as a field in which a small magnet of unit moment experiences a couple of unit magnitude when held with its magnetic axis at right angles to the direction of the field. A small magnetic needle, therefore, sets itself, when free to move in a magnetic field, in a direction which indicates the direction of this field at that point. If the needle is forced round into a position at right angles to the direction of the field, then the couple required to hold the needle in this new position is a measure of the strength of the field at that point.

If two magnetic fields are created in one region which are in directions at right angles to each other, and if a small magnet is held in this region, it will take up a direction which indicates the *resultant field* at that point. A magnetic field has, therefore, direction and magnitude, and may be indicated on a diagram by a straight line, like a force or a velocity. Magnetic fields are combined and resolved in accordance with the rules for the resolution and composition of forces or velocities.

If H is the magnitude of one field (represented by a straight line), and F is the magnitude of another field at right angles to H, and if a small exploring needle is held in the combined field, the direction of the needle will be such that its axis makes an angle θ with the direction of the field H, so that

$$\frac{F}{H} = \tan \theta, \text{ or } F = H \tan \theta.$$

If, therefore, H has a constant value, and F is variable in magnitude, we can find the relative values of F at various points by observing the tangents of the angles of deviation of a small exploring magnet placed in that region at those points.

Pass a constant electric current through the circular coil of wire, and at any point on the axis of this coil place the small compass needle. Turn the coil round until the direction of the axis of the coil is at right angles to the magnetic meridian at that point. Observe the angle of deviation θ of the small magnetic needle when placed at different distances D from the centre of this coil. Find out from a table of tangents the tangents of these angles, and measure at the same time the distance from the centre of the coil to the centre of the small needle. Enter up your results in Table I. opposite. Plot a curve showing the decrease of magnetic force due to the coil at various points along the axis of the coil. If the field of the coil is very strong relatively to that of the earth, use a small controlling magnet to increase the strength of the constant field.

At any point P taken on a line drawn through the centre of a short magnet of length l and at right angles to its axis the magnetic force at that point is parallel to the axis of the magnet, and has a value which varies inversely as the cube of the distance from either pole of the magnet.

Draw a diagram of magnetic forces proving this as a consequence of the fact that the force at any point due to *each* pole varies as $\frac{m}{r^2}$, where m is the strength of the pole and r the distance of the point. Prove it experimentally by placing a *short* magnet with its axis at right angles to the magnetic meridian, and by holding a small compass needle at various points on a meridian drawn through the centre of the *short* magnet. Observe the angles of deflection of the needle. Enter up your results in Table II. Show that the product of the cube of the distance of the centre of the needle from the centre of the magnet and the tangent of the angle of deflection is a constant quantity. Draw a curve representing this decreasing force at various points on the equatorial line of a short magnet.

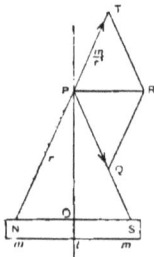

If N and S are the poles of the magnet (*see* diagram), and if $NS = l$, and if the point P is taken on a line OP drawn through the centre O of the magnet at right angles to NS, then the force at P due to each pole of strength m is $\frac{m}{r^2}$, where $r = NP$ or SP. From the similarity of the triangles P T R, N P S, it will be seen that the resultant P R of these two forces $\frac{m}{r^2}$ has a magnitude $\frac{m \, l}{r^2 \, r} = \frac{M}{r^3}$, since $P R : \frac{m}{r^2} = l : r$. Hence the resultant force varies inversely as the cube of r, that is, nearly as the cube of the distance from the point P to the centre of the short magnet.

THE EXPLORATION OF MAGNETIC FIELDS.

TABLE I.—*The Axial Field of a Circular Current.*

Observation No.	$\theta =$ Deflection of small exploring needle.	Tan θ.	D = Distance of centre of exploring needle from centre of coil.	H = Constant magnetic field due to earth or other magnets.	F = H tan θ. Magnetic force at distance D from centre of coil.

THE EXPLORATION OF MAGNETIC FIELDS.

TABLE II.—*The Equatorial Field of a Short Magnet.*

Observation No.	$\theta =$ Deflection of small exploring needle.	Tan θ.	D = Distance of centre of exploring needle from centre of short magnet.	H = Constant magnetic field due to earth.	$D^3 \tan \theta = \frac{M}{H}$ = ratio of moment of short magnet to field of earth.

ELECTRICAL LABORATORY NOTES AND FORMS.

No. 2.—ELEMENTARY.

Name.....................   *Date*_____

The Magnetic Field of a Circular Current.

The apparatus required for these experiments is a galvanometer consisting of one turn of a thick wire bent into a circle, and a magnetic compass box with a degree scale, arranged to slide along a bar so that the centre of the needle is always in a line drawn through the centre of the circular current, and in a direction perpendicular to its plane.

The Student is recommended to sketch the arrangement of the apparatus.

If a conductor, such as a wire, is bent into a circle of one or more turns and an electric current sent through it, a magnetic field is created all around this conductor, which is called *the field of the current.* Pass a current of four or five amperes through such a circular coil, and observe the general direction of the field at various points by means of iron filings. At points on a line (called *the axis of the circle*) drawn perpendicularly to the plane of the coil and passing through its centre, the magnetic force is in the direction of this axis, and has various decreasing values as the point is taken farther from the plane of the coil.

If a circle of a single turn of wire of mean radius R has a current of A amperes passed through it, the magnetic force at the centre is equal to $\frac{2\pi}{10} \frac{R\,A}{R^2}$ units, being proportional to the circumference of the circle and inversely as the square of its radius. This force is perpendicular to the plane of the circle. If the coil has N turns, the magnetic force at the centre of the coil is equal to

$$\frac{2\pi}{10}\ \frac{N\,A}{R} = \frac{5}{8}\ \frac{\text{ampere turns}}{\text{mean radius}}\ \text{units.}$$

If the point is taken along the axis of a circular coil of mean radius R and of N turns at a distance D from the centre, the resultant magnetic force at this point due to a current of A amperes in this coil is equal to F, where

$$F = \frac{2\pi}{10}\ \frac{R\,N\,A}{(R^2 + D^2)}\ \frac{R}{\sqrt{R^2 + D^2}}\ \text{units} = \frac{5}{8}\ N\,A\ \frac{R^2}{(R^2 + D^2)^{\frac{3}{2}}}\ \text{units.} \qquad (1)$$

Remember that the magnetic force at a point means the force *in dynes* which would act on a unit magnetic pole held at that point. Hence, if a *small* magnet of length *l*, and having poles of strength *m*, is placed with its length parallel to the plane of the coil at any point of the axis of the circle, the *couple* or *torque* tending to twist this magnet round will be equal to $F\,m\,l$, or to

$$\frac{5}{8}\,N\,A\,\frac{R^2}{(R^2+D^2)^{\frac{3}{2}}}\,m\,l \text{ units, or to}$$

$$\frac{5}{8}\,\frac{\text{square of radius of coil} \times \text{ampere-turns} \times \text{moment of magnet}}{\text{cube of mean distance from magnet to circumference of circle}} \quad . \quad (2)$$

Note that the product $m\,l = M$ is called *the moment of the magnet.*

This couple tends to set the magnet with its length perpendicular to the plane of the circle.

If another constant magnetic force of value H is made to act in a direction perpendicular to the axis of the circular coil, under these two forces, the small needle will take a position with its length inclined at an angle θ to the plane of the circular coil, and the ratio of F to H is equal to the tangent of θ, or $F = H \tan \theta$. Equating the values for the magnetic force F, we have

$$H \tan \theta = \frac{5}{8}\,N\,A\,\frac{R^2}{(R^2+D^2)^{\frac{3}{2}}} \quad . \quad . \quad . \quad . \quad (3)$$

Hence, for a given circular coil and a constant current passed through it, the magnetic force at points on the axis varies inversely as the cube of the distance of the axial point from the mean circumference of the coil, and the magnetic force varies as the tangent of the angle of deflection of the small needle held at that point. Prove the above formula (3) by using the sliding magnetometer and coil, consisting of a circular coil of insulated wire and a small magnetic needle which can be placed at various points on the axis of the coil. Put the needle at various distances from the plane of the coil and observe the angles of deflection of the needle due to a constant current in the coil and a constant controlling magnetic force H. Try also different values of the current A, and use the formula (3) either to find the value of A, knowing H and the dimensions of the coil; or to find H, knowing A and the value of the dimensions R, D and N, as given to you, for the particular coil you are using.

Plot a curve showing the varying values of the magnetic force along the axis of a circular coil at points taken equidistantly along this axis.

Enter up your observations in the form given on the opposite page.

Attend to the following points :—In setting up the magnetometer, place it so that the direction of the earth's horizontal magnetic field (equal to 0·18 of a unit) is in a direction parallel to the plane of the coil. See that the current in the leading-in wires or resistances does not directly disturb this needle. Prove this by short-circuiting the circular coil. Take care to keep the current very constant in the circular coil by using an adjustable resistance (carbon) and an ammeter in series with it. The constants for the coil you are using are as follows :

R =   c.m.   N =

THE MAGNETIC FIELD OF A CIRCULAR CURRENT.

Observation No.	$\theta =$ Angle of deflection of needle.	$\tan \theta =$ Tangent of angle of deflection.	D = Distance of needle from centre of coil. =	A = Ampere current in coil.	H = Magnetic controlling force.	$y =$ Calculated value of $\tan^2 \theta \ (R^2 + D^2)^3$.

THE MAGNETIC FIELD OF A CIRCULAR CURRENT.

Observation No.	$\theta =$ Angle of deflection of needle.	$\tan \theta =$ Tangent of angle of deflection.	D = Distance of needle from centre of coil.	A = Ampere current in coil.	H = Magnetic controlling force.	$y =$ Calculated value of $\tan^2 \theta (R^2 + D^2)^{3/2}$.

ELECTRICAL LABORATORY NOTES AND FORMS.

No. 3.—ELEMENTARY.

Name   *Date*

The Standardization of a Tangent Galvanometer by the Water Voltameter.

The apparatus required for these experiments is a small tangent galvanometer, and a water voltameter (preferably Hofmann's form), in which the two gases are separately collected in two tubes connected by a cross tube and to a reservoir of dilute acid. Use 1 : 10 dilute sulphuric acid as the electrolyte.

The Student is recommended to sketch the arrangement of the apparatus.

When an electric current is passed through an *electrolyte*, or electrically decomposable liquid contained in a *voltameter*, it extricates from it two ingredients, called the *ions*, and delivers these at the metallic or other plates by which the current leaves and enters the liquid, and which are called the *electrodes*. The metal in the solution, or the equivalent body, such as hydrogen, is delivered up at the negative electrode. The amount in grammes of this metal or ion so delivered up in one second is called its *electro-chemical equivalent*. The practical definition of a current of *one ampere* is that it is an unvarying current which deposits in one second ·001118 gramme of silver from a dilute solution of nitrate of silver. A current of one ampere deposits in one hour at the negative electrode (when passed through solutions of these salts) 4·025 grammes of silver, 1·178 grammes of copper, and ·03738 gramme of hydrogen. These are called the electro-chemical equivalents per ampere-hour. It can be shown that ·03738 gramme of hydrogen has a volume of very nearly 440 cubic centimetres (440·15 exactly) when measured at 15°C. and at a barometric pressure of 760mm. If an electric current is sent through a dilute solution of sulphuric acid (1 vol. acid, 9 vols. water), and it liberates at the negative pole a volume of v cubic centimetres of hydrogen, measured at t°C. and H mm. barometric pressure, in x seconds, we can calculate what this volume v would become at the standard pressure of 760mm. by multiplying v by $\dfrac{H}{760}$. Similarly, we can reduce it to the standard temperature (15°C.) by multiplying again by $\dfrac{273+15}{273+t}$.

In practice, the resistances P and Q are the two sections of a long fine wire stretched on a scale, and S is a plug resistance box. A battery of one or two cells is

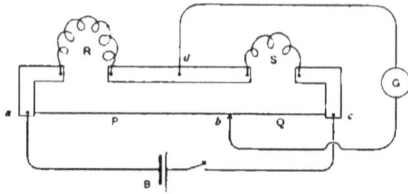

joined to the ends of the wire, and the rectangle is completed by joining the standard resistance S and the unknown resistance R as in the second diagram. The galvanometer is joined in between the junction of S and R and the sliding contact on the bridge wire. Having joined up the resistances as above, find the point of contact of the slider on the wire at which the galvanometer gives no deflection when the battery key is down.

Attend to the following points :—

(*i.*) Put down the battery key first before making contact on the slide wire.

(*ii.*) Press the slide wire key as lightly as possible so as not to nick the slide wire.

(*iii.*) Obtain the balance as quickly as possible, because the current heats the conductors and therefore alters the resistance to be measured.

When balance is obtained, read off the length of the two sections of the divided wire on either side of the slider. These lengths are proportional to the resistances of the sections, which are the resistances P and Q. Note the standard resistance S used, and calculate the value of the unknown resistance R from the equation

$$R = \frac{P}{Q} S.$$

P and Q are called the ratio arms of the bridge. The accuracy of the measurement depends, therefore, on the uniformity of the slide wire. Test this by measuring a few known resistances. The most favourable arrangement for sensitiveness is when the standard resistance S has a value about equal to the resistance R to be measured.

Enter up your results in the form given on next page. Note that the sign ω stands for ohms, and mm. for millimetres.

In order to assist in finding approximately the position of balance on the slide wire it is convenient to shunt the galvanometer by a German silver wire, which is removed when the balance is nearly obtained. This shunt wire should have about one-tenth the resistance of the galvanometer itself.

THE MEASUREMENT OF ELECTRICAL RESISTANCE BY THE DIVIDED WIRE BRIDGE.

Observation No.	Ratio arms of bridge in mm.		Standard resistance in ohms. S =	Calculated value of unknown resistance in ohms. R =	Description of wire or conductor measured.
	P =	Q =			

THE MEASUREMENT OF ELECTRICAL RESISTANCE BY THE DIVIDED WIRE BRIDGE.

Observation No.	Ratio arms of bridge in mm.		Standard resistance in ohms. S =	Calculated value of unknown resistance in ohms. R =	Description of wire or conductor measured.
	P =	Q =			

ELECTRICAL LABORATORY NOTES AND FORMS.

No. 5.—ELEMENTARY.

Name.. *Date*....................................

The Calibration of the Ballistic Galvanometer.

*The apparatus required for these experiments is a sensitive mirror ballistic galva-
nometer of suspended coil or suspended needle type, a condenser of variable
capacity, a few secondary cells, and a voltmeter to measure potential. A
discharge key is also required.*

The Student is recommended to sketch the arrangement of the apparatus and circuits.

A galvanometer consists in general of a magnet and a coil of wire. The magnet
may be fixed and the coil suspended, or the coil fixed and the magnet suspended. In the
normal position the magnetic axis of the magnet is at right angles to the magnetic axis
of the coil. When a current is passed through the coil, the coil and the magnet act on
one another in such a manner that they tend to place their magnetic axes parallel
to one another. This movement is resisted by another force, which is called the *control*.
The control in the case of the movable magnet galvanometer is generally an external
magnet creating a constant field. The control in the case of the movable coil galvanometer is generally the torsion of a wire or other suspension. If means are taken to
bring the movable part, whether coil or magnet, to rest when disturbed as quickly as
possible, this arrangement is called a *damped* galvanometer. If the arrangements are
such as to permit the movable part to execute vibrations when disturbed, which are as
little *damped* or resisted as possible, then it is called a *ballistic galvanometer*. The
use of a ballistic galvanometer is to measure the quantity of electricity. The use
of a damped galvanometer is to measure steady current strength.

When any heavy body is movable round an axis, the sum of the products
obtained by multiplying the mass of each particle m of the body by the square of its
distance r from the axis, or taking the sum $\Sigma\, m\, r^2$, is called its *moment of inertia*
round this axis, and this is denoted by the letter I.

In the case of bodies swinging round an axis the following are important
quantities :—

> The *angular velocity* $= \omega = \dfrac{d\theta}{dt}$, or the ratio of the numerical values of the
> small angle $d\theta$ described to that of the small time $d\,t$ occupied in
> describing it.

> The *angular momentum* $= \mathrm{I}\,\omega = $ product of moment of inertia and angular
> velocity at any instant.

> The *angular energy* $= \frac{1}{2}\,\mathrm{I}\,\omega^2 = $ product of $\frac{1}{2}$ square of angular velocity
> and moment of inertia.

> The *angular force* or *couple* or *torque* which is causing rotation $= \mathrm{T} = \mathrm{I}\,\dfrac{d\omega}{dt}$,
> rate of change of angular momentum.

In the case of a small magnet, of length l, vibrating round a vertical axis and set in vibration by the application of a magnetic force, which acts always at right angles to the length of the magnet, and has a magnitude f at any instant, the *couple* acting is $f\,l$, and by the above equation

$$f\,l\,d\,t = I\,d\,\omega. \quad \ldots \ldots \ldots \ldots (1)$$

Note that the product of the magnitude of a couple and the time during which it acts is called the *impulse of the couple*. Hence, if a couple of varying magnitude acts as above on a magnet, and if during the impulse the force on the magnetic poles always acts at right angles to the magnet's axis, and if this impulse gives the magnet a "throw" or angular deviation θ, such that it starts off from its position of rest with an angular velocity Ω, it is seen from equation (1) that the total impulse acting on the needle is measured by the total angular momentum with which the needle starts off.

If the magnet hanging in a galvanometer has a magnetic moment M, and if a short electric current or discharge is sent through the galvanometer coils, and if i is the strength of this current *at any instant*, the value of i is obtained as the numerical ratio of $d\,q$ to $d\,t$, where $d\,q$ is the small quantity of electricity which flows through the coils in the small time $d\,t$. Hence $i\,d\,t = d\,q$. If the whole of the discharge is over before the magnet can move from its initial position, the impulse of the couple acting on it is the integral or sum of all the small impulses, $M\,C\,i\,d\,t = M\,C\,d\,q$, where C is a constant depending on the form of the coils. Hence the total impulse of the couple acting on it is equal to $M\,C\,Q$, where Q is the whole quantity of the discharge. This, by the above principles, is numerically equal to $I\,\Omega$, where Ω is the angular velocity with which the needle starts off from its position of rest.

The kinetic energy of the needle at starting is then measured by $\frac{1}{2}I\,\Omega^2$. The potential energy of the needle when just at rest for an instant at the extremity of its swing θ is $M\,H\,(1-\cos\theta) = 2\,M\,H\,\sin^2\frac{\theta}{2}$, where H is the magnitude of the controlling field. Hence,

$$\tfrac{1}{2}I\,\Omega^2 = 2\,M\,H\,\sin^2\frac{\theta}{2},\text{ and }I\,\Omega = M\,C\,Q;$$

therefore,
$$Q = 2\sqrt{\frac{H\,I}{M\,C^2}}\sin\frac{\theta}{2}. \quad \ldots \ldots \ldots (2)$$

If, therefore, the discharge is all over before the needle has time to move from its position of rest, the total quantity of electricity in the discharge Q varies as the sine of half the angle of "throw" produced by it.

Prove this for the mirror ballistic galvanometer by charging a condenser of capacity K to various potentials V, and discharging this quantity $Q = K\,V$ through the ballistic mirror galvanometer, noting the excursion x of the spot of light. If d is the distance of scale from mirror, obtain from d and x the value of $\sin\frac{\theta}{2}$. Note that $x = d\tan 2\,\theta$. Record your results in the form on next page.

Note that owing to the damping action of the air the observed $\sin\frac{\theta}{2}$ should be multiplied by a factor $\left(1+\frac{\lambda}{2}\right)$ to obtain the true value which $\sin\frac{\theta}{2}$ would have if no retardation existed. The value of λ is obtained by taking the common logarithm of the ratio of the excursion x_1 of one swing to that of the next—viz., x_2—when the needle is vibrating freely, and multiplying this logarithm by 2·3026. λ is called *the logarithmic decrement* of the galvanometer.

THE CALIBRATION OF THE BALLISTIC GALVANOMETER.

Observation No.	$d =$ Distance of scale from mirror.	$x =$ Excursion of spot of light.	Tan $2\,\theta$ $= \dfrac{x}{d}$	Corrected value of $\sin\dfrac{\theta}{2}$	K = Capacity of condenser.	V = Potential of charge.	Value of $\dfrac{K\,V}{\sin\dfrac{\theta}{2}}$ = ballistic constant.

THE CALIBRATION OF THE BALLISTIC GALVANOMETER.

Observation No.	$d =$ Distance of scale from mirror.	$x =$ Excursion of spot of light.	Tan 2θ $= \dfrac{x}{d}$	Corrected value of $\sin \dfrac{\theta}{2}$	$K =$ Capacity of condenser.	$V =$ Potential of charge.	Value of $\dfrac{K V}{\sin \dfrac{\theta}{2}}$ $=$ ballistic constant.

ELECTRICAL LABORATORY NOTES AND FORMS.

No. 6.—ELEMENTARY.

Name   *Date*_____

The Determination of Magnetic Field Strength.

The apparatus required for these experiments is a mirror ballistic galvanometer having a needle or coil with time of oscillation not less than two or three seconds and a known small or negligible logarithmic decrement, a box of resistance coils, a suitable exploring loop of wire, and a magnet.

The Student is recommended to sketch the arrangement of the apparatus.

If a loop of one or more turns of insulated wire is held in the magnetic field of a magnet, or of an electric conductor conveying a current in such a position that some of of the lines of induction of the magnet or current pass through the loop, this circuit is said to be *linked* with the lines of induction. If this loop of wire is connected with a ballistic galvanometer (previously calibrated), and is suddenly snatched away from its position, these lines of induction are all taken out of the loop and an electromotive force thus created which sets in motion a certain quantity of electricity and sends it through the galvanometer. Under these conditions the "throw" of the needle of the ballistic galvanometer becomes a measure of the number of lines of induction passing through the loop, or of the magnetic field strength at that point. Let H stand for the strength of the magnetic field at any point ; that is, for the number of lines of induction which pass through one square centimetre of area of the loop of wire when it is held in the field, so that its plane is perpendicular to the direction of the field at that point. Let N be the number of turns of wire on the loop, and A the mean area of each turn, in square centimetres or fractions of a square centimetre.

The product H N A is called the *total induction* through the loop, and is denoted by B. If the loop is taken away to a place where there is no sensible magnetic field, the value of B varies from B to zero. At any instant the electromotive force set up in the loop is measured by the rate at which B is changing at that instant. If the loop is connected with a ballistic galvanometer, and if R is the resistance in ohms of the whole circuit consisting of the loop, galvanometer, and connecting wires, then the current which is flowing in this circuit *at any instant* is measured by the quotient of the electromotive force in the circuit at that instant, by the total resistance of the circuit, or by the quotient of the rate of change of the total induction through the circuit at that instant by the total resistance of the circuit. If we divide the whole time during which the loop is being moved away from the field into little elements of time *d t*, during each of which the current in the loop circuit is i, the sum of all the products of the quantities i and dt from the beginning to the end of the movement is the measure of the total quantity of electricity, Q, which has been set flowing through the loop circuit. Hence, the quotient obtained by dividing the total change in induction through the circuit by the total resistance of the circuit, is numerically equal to the total quantity of electricity which has flowed through the circuit, or,

$$\frac{B}{R} = Q, \text{ or } B = R\,Q,$$

or
$$H N A = R Q \quad . \quad . \quad . \quad . \quad . \quad . \quad . \quad (1)$$

Hence, to obtain the value of the field H at any point, we have to hold a loop of wire of N turns, each turn of the loop having an area A, at that point in such a manner that the lines of induction pass perpendicularly through the loop. We have then to connect this loop with a ballistic galvanometer, and measure the resistance R of the whole circuit. If, then, the loop is snatched away, the ballistic galvanometer needle will make a sudden deflection or "throw" θ, and the value of $C \sin \frac{\theta}{2}$ is a measure of this total quantity of electricity which has been sent through the galvanometer. Hence

$$Q = C \sin \frac{\theta}{2},$$

where C is the ballistic constant of the galvanometer (*see* Elementary Form No. 5). We have then

$$H N A = R C \sin \frac{\theta}{2},$$

or
$$H = \frac{R C \sin \frac{\theta}{2}}{N A}. \quad . \quad \quad . \quad . \quad . \quad . \quad (2)$$

The value of $\sin \frac{\theta}{2}$ is easily found when we know the excursion x which the spot of light makes (assuming a mirror galvanometer used) and the distance d of the scale from the mirror, for $\frac{x}{d} = \tan 2\,\theta$, and hence θ, and therefore $\sin \frac{\theta}{2}$, can be found.

Note the following points. The correctness of the above reasoning depends upon the assumption that the ballistic galvanometer has a needle with such a slow period of swing that the whole of the impulse on the needle is over and complete before the needle has time to move from its normal position. Hence the loop must be snatched away very quickly. The needle must be quite stationary before taking the "throw." Using a ballistic galvanometer, take such "throws" with a loop circuit held at various points in the field of a magnet, and obtain numbers which are proportional to the relative strengths of the field at those different points. Take as the unit field of comparison the strength at some fixed point. Explore thus the field at various points along the axis of a magnet, using the field strength right up against the pole as a standard of comparison. Before entering the values of $\sin \frac{\theta}{2}$ in the table correct all the observed values of $\sin \frac{\theta}{2}$ by multiplication by the factor $\left(1 + \frac{\lambda}{2}\right)$, where λ is the logarithmic decrement of the galvanometer.

THE DETERMINATION OF MAGNETIC FIELD STRENGTH.

Observation No.	x Excursion of spot of light.	d Distance of scale from mirror.	$\dfrac{x}{d}$ $= \tan 2\theta$.	Corrected value of $\sin \dfrac{\theta}{2}$ $= S$	$S^1 =$ Corrected value of $\sin \dfrac{\theta}{2}$ for Standard position of loop.	Ratio of $\dfrac{S}{S^1}$.

THE DETERMINATION OF MAGNETIC FIELD STRENGTH.

Observation No.	x Excursion of spot of light.	d Distance of scale from mirror.	$\dfrac{x}{d}$ $= \tan 2\,\theta$.	Corrected value of $\sin\dfrac{\theta}{2}$ $= S$	$S^1 =$ Corrected value of $\sin\dfrac{\theta}{2}$ for Standard position of loop.	Ratio of $\dfrac{S}{S^1}$.

ELECTRICAL LABORATORY NOTES AND FORMS.

No. 7.—ELEMENTARY.

Name   *Date*......

Experiments with Standard Magnetic Fields.

The apparatus required for the following experiments is:—A straight coil or long bobbin of insulated wire wound on a hollow pasteboard, glass, or wooden tube; the total number of turns of the wire being known, and the distance between the cheeks of the coil. Also a carefully-wound bobbin of wire with one layer of windings, the total area included by all the windings being known. A sensitive mirror ballistic galvanometer, a box of resistance coils, and an exploring coil or loop are needed.

The Student is recommended to sketch on foolscap-sized paper the arrangement of the apparatus and the connections.

The magnetic force or field strength F at a point in the neighbourhood of a very long straight wire carrying a current of A amperes is numerically equal to the quotient of one-fifth of the current by the distance d of the point from the wire in centimetres, or

$$F = \frac{2A}{10\,d}. \tag{1}$$

This formula is true only if the return conductor is at a very great distance. If a circle of radius d is described round the wire, with its centre on the axis of the wire and its plane perpendicular to the wire, this circle is a line of force; since the magnetic force at every point is in the direction of this circle, and has a magnitude equal to that given by equation (1). The length of this line is equal to $2\pi d$. This line is called also a magnetic circuit.

If we multiply together the length of this magnetic circuit, viz., $2\pi d$, by the value of the magnetic force along that circuit, viz., $\frac{2A}{10\,d}$, we obtain a product, viz., $\frac{4\pi}{10}A$, which is called the *magneto-motive force* along that line. In any case, whatever may be the form of the conducting circuit, or of the magnetic circuit linked with it, we always have the following general relation, viz. :

$$\left\{\begin{array}{l}\text{The magneto-motive force along a}\\ \text{magnetic circuit}\end{array}\right\} = \frac{4\pi}{10} \times \left\{\begin{array}{l}\text{The total ampere-current flowing through,}\\ \text{or linked with, that magnetic circuit.}\end{array}\right\}$$

Apply this to a general case. Let a wooden ring, of which the mean diameter is large compared with the diameter of its circular cross section, be wound closely over with insulated wire. Let there be N turns of wire. Let a current of A amperes be sent through the wire. This forms what is called a circular solenoid. The magnetic force in the axis of this solenoid is everywhere along the circular axis of the solenoid. Call the length of this circular axis L, and the magnetic force F. The total current flowing through this magnetic circuit is NA, and, by the above rule,

$$F L = \frac{4\pi}{10} N A.$$

Hence
$$F = \frac{4\pi}{10} \cdot \frac{N A}{L} ; \qquad \qquad . \quad (2)$$

or the magnetic force in the centre of the circular solenoid is numerically equal to $\frac{4\pi}{10}$ times the ampere-turns per unit of length of the solenoid. The above rule also holds good for a straight solenoid or bobbin of wire, provided its length is large compared with its diameter. Hence the magnetic force in the interior of a long bobbin of insulated wire is obtained by multiplying $\frac{4\pi}{10}$ by the ampere-turns on the bobbin per unit of length. Hence a long bobbin of this kind provides us with a standard magnetic field when a known ampere-current is passed through it.

The dimensions of the long bobbin of wire given to you are as follows :—

Length between the cheeks   =   centimetres.
Total number of turns of wire =

Calculate by equation (2) the magnetic force in the centre of the bobbin when currents of 1, 2, 3, 4, and 5 amperes respectively are sent through the wire.

Place the other or secondary bobbin of wire in this known field, and connect it through a plug resistance box with the ballistic galvanometer. Measure the whole resistance of the circuit composed of the galvanometer secondary coil and connections. Suddenly interrupt the current through the standard coil and note the "throw" of the ballistic galvanometer. Try this with a constant current passing through the standard coil and various resistances unplugged out of the resistance box. Prove that the product of the sine of half the angle of throw of the galvanometer needle, or $\sin \frac{\theta}{2}$, and the total resistance of the galvanometer circuit is a constant quantity. Correct all the observed values of $\sin \frac{\theta}{2}$ by multiplying by $\left(1 + \frac{\lambda}{2}\right)$, where λ is the *logarithmic decrement* of the galvanometer. (*See* Elementary Form No. 5.)

Enter up your results in Table I.

Next take the standard secondary coil, which consists of a single layer of fine silk-covered wire wound uniformly on a tube of known diameter, the total area included by all the wire windings being known.

Place this secondary bobbin in the known field of the standard coil, and pass various currents through this latter. Observe the throw θ of the needle of the ballistic galvanometer when this current is arrested, and prove that $\sin \frac{\theta}{2}$ varies as the field of the standard coil when the resistance of the galvanometer circuit remains constant ; in other words, prove that the quotient of $\sin \frac{\theta}{2}$ by the ampere-current of the standard coil is constant.

Enter up your results in Table II.

Note that in all these experiments either a suspended coil ballistic galvanometer must be used, or else the standard coil must be placed so far away from the galvanometer as not to affect it magnetically.

EXPERIMENTS WITH STANDARD MAGNETIC FIELDS.

TABLE I.

Observation No.	$d =$ Distance of mirror from scale.	$x =$ Excursion of spot of light.	Tan 2θ $= \dfrac{x}{d}$	Corrected value of $\sin \dfrac{\theta}{2}$.	$R =$ Total resistance of galvanometer circuit.	Calculated and corrected value of $R \sin \dfrac{\theta}{2}$.

EXPERIMENTS WITH STANDARD MAGNETIC FIELDS.

TABLE II.

Observation No.	$d =$ Distance of mirror from scale.	$x =$ Excursion of spot of light.	Tan 2θ $= \dfrac{x}{d}$.	Corrected value of $\sin \dfrac{\theta}{2}$.	$A =$ Current through standard coil.	Calculated and corrected value of $\dfrac{\sin \dfrac{\theta}{2}}{A}$.

ELECTRICAL LABORATORY NOTES AND FORMS.

No. 8.—ELEMENTARY.

Name   *Date*

The Determination of the Magnetic Field in the Air Gap of an Electromagnet.

The apparatus required for the following experiments is a ballistic galvanometer and condenser for standardising it, an exploring coil consisting of a small flat bobbin wound over with a large number of turns of fine insulated wire, and an electromagnet consisting of two soft iron bars, each bent into a semicircle or half rectangle, and having on each middle part a magnetising coil of a known number of turns.

The Student is recommended to sketch the arrangement of the circuits.

If any magnetic circuit, whether consisting wholly of iron, wholly of non-magnetic material, or partly of one and partly of the other, is subjected to a magnetising force, this force produces *magnetic induction* in the iron or other material. If the circuit is wholly of iron in the form of a ring and the magnetising force applied by means of a current sent through a coil wound on the iron ring, the magnetising force can be calculated by a known formula; (*see* equation (2), Elementary Form No. 7). If the magnetic circuit consists of an iron ring or circuit having a narrow air gap or cut made perpendicularly across it, then the magnetic force in this gap is called *the induction* in the iron. If a small loop of wire of one or more turns is held in this air gap, or looped round the iron circuit, and then pulled suddenly away from the field, an induced electromotive force is set up in the loop. If the loop is connected with a ballistic galvanometer this transitory electromotive force creates an electric flow through the galvanometer. The quantity, Q, of electricity sent through the ballistic galvanometer when the loop is pulled away from the field can be determined when we know the sine of half the angle of deflection θ of the needle produced by it, and the ballistic constant C of the galvanometer (*see* Elementary Form No. 5), since

$$Q = C \sin \frac{\theta}{2} \left(1 + \frac{\lambda}{2} \right),$$

where λ is the logarithmic decrement of the galvanometer.

If R is the resistance of the exploring coil, connections, and galvanometer measured in ohms, then, when the loop is placed in the air gap so that the lines of induction pass at right angles through it and it is pulled suddenly away, the relation between the quantity of electricity Q set flowing, the total resistance of the galvanometer circuit R, the induction in the air gap B, the number of turns on the exploring coil, N and the mean area of each turn of that coil A, is as follows (*see* Elementary Form No. 6) :—

$$A N B = Q R \quad . \quad . \quad . \quad (1)$$

The quantity A N B is called *the total induction* through the galvanometer loop, and is the product of the mean area of each turn of the loop, the number of turns on the loop, and the mean induction at the place where the loop is placed. Hence B can be determined in centimetre-gramme-second (C.G.S.) units when the quantities R, A, N, Q are known in the same units.

Note that, if R is measured in ohms, the number must be *multiplied* by 10^9 to reduce it to C.G.S. units, and if Q is measured in microcoulombs that number must be *divided* by 10^7 to reduce it to C.G.S. units, and then those last values used in equation (1), A being also measured in square centimetres. The values of A and N for the small exploring coil given to you are as follows :—

$$A = \qquad\qquad \text{square centimetres.}$$
$$N =$$

Take the exploring loop and connect it in series with a ballistic galvanometer, and measure the resistance R of the whole circuit. Place this loop in the air gap of an electromagnet having a known number of turns n on its exciting coil. Pass various exciting currents a amperes through this electromagnet coil, and note the ampere-turns $a\,n$ employed as excitation. Place the small coil in the air gap, and pull it suddenly away. Observe the throw of the ballistic galvanometer θ, and the logarithmic decrement λ of the same. Standardise the galvanometer by the condenser, and determine the ballistic constant C (*see* Elementary Form No. 5) to reduce the values of $\sin \frac{\theta}{2}\left(1+\frac{\lambda}{2}\right)$ to electric quantity Q in microcoulombs, and calculate the value of the induction B in the air gap from equation (1). Plot a curve showing the relation of the ampere-turns of excitation to the induction in the iron of the electromagnet. Enter up the results of your observations in the Table opposite. Try the same experiments, only keeping the ampere-turns of exciting current constant and varying the air gap width. Plot a curve showing the variation of induction with air gap length. Try these same experiments with a small model dynamo having armature cores of various clearances to vary the air gap width, taking a series of observations with different exciting currents for each air gap width, and so determine the induction across the air gap.

THE DETERMINATION OF THE MAGNETIC FIELD IN THE AIR GAP OF AN ELECTROMAGNET.

The following are the constant values required :—

A = mean area of exploring coil   =   centimetres.
N = number of turns on exploring coil   =
n = number of turns on field magnet   =
d = distance of galvanometer needle from scale =   centimetres.
R = total resistance of galvanometer circuit   =   ohms.
λ = logarithmic decrement of galvanometer   =
C = ballistic constant of galvanometer   =

Calculate B from the formula

$$B = \frac{C \sin \frac{\theta}{2}\left(1 + \frac{\lambda}{2}\right) R \cdot 10^{\rho}}{A N \cdot 10^{7}}.$$

Observation No............	Ampere current in field coils $= a$.	Ampere turns of field magnet $= a\,n$.	Throw of galvanometer spot of light in centimetres $= r$ $\dfrac{r}{d} = \tan 2\,\theta$.	$\sin \frac{\theta}{2}\left(1 + \frac{\lambda}{2}\right)$.	$w =$ Width of air gap.	Induction in air gap $= B$.

THE DETERMINATION OF THE MAGNETIC FIELD IN THE AIR GAP OF AN ELECTROMAGNET.

Observation No............	Ampere current in field coils $= a$.	Ampere turns of field magnet $= a\,n$.	Throw of galvanometer spot of light in centimetres $= x$. $\dfrac{x}{d} = \tan 2\,\theta$.	$\operatorname{Sin}\dfrac{\theta}{2}\left(1+\dfrac{\lambda}{2}\right)$.	$w =$ Width of air gap.	Induction in air gap $= $ B.

No. 9.—ELEMENTARY.

Name   *Date*

The Determination of Resistance with the Post Office Pattern Wheatstone Bridge.

The apparatus needed for these experiments is a Post Office pattern of "Wheatstone's Bridge," a suspended coil galvanometer, a few cells of a dry battery, some coils of wire for experiment, a vessel of paraffin oil in which to heat the coils, and a thermometer.

The Student is recommended to sketch the arrangement of the circuits.

The arrangement of the coils in the Post Office pattern of Wheatstone's Bridge is shown in the annexed diagram. The top row of coils is called the ratio arms, and is generally a series of six coils, 1,000, 100, 10, 10, 100, 1,000 ohms; in large bridges 10,000 ohms and 1 ohm being also added. The balancing coils consist of a series of coils from 4,000 or 5,000 ohms to 1 ohm. The galvanometer is generally connected (through a key) with the extremities of the ratio arm series of coils, and the battery (through a key) with the centre of this series and the end of the series of balancing coils. With regard to the plugs, note the following precautions :—

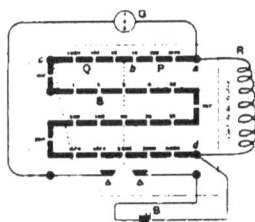

(*i.*) Do not force in the plugs too hard, or the heads will be twisted off in getting them out.

(*ii.*) Put in the plugs firmly, however, when in use, so as to avoid any loose contacts.

(*iii.*) Never touch the brass part of the plug with grease or dirty fingers.

(*iv.*) Do not leave the plugs that are not in use on the table, but keep them in the lid of the bridge box.

(*v.*) When you have finished with the bridge replace all the plugs *loosely* in their holes and close the bridge box.

To make a measurement of resistance, take the coil of wire given to you and connect it by thick copper wires with the bridge, arranging as shown in the diagram. See that all connections are tight and good. Place the coil to be measured in the paraffin oil bath, and take the temperature of the oil, stirring it well. Remove a couple of the plugs in the ratio arms, of equal value, say 100 and 100 ; then try taking out various plugs from the balancing series until a balance is obtained, or a condition found such that, with plug 1 out, the small deflection of the galvanometer is one way, and with plug 1 in it is the other way. Then try another ratio, taking out, say, 1,000 on one side and 10 on the other, in the ratio arm series, the 10 being on side nearest to the coil being measured, and try again.

When either an exact balance has been obtained, or else a deflection one way with plug 1 out and the other way with plug 1 in, the correct resistance may be obtained as follows :—

Suppose the ratio arms are 1,000 : 10 and the plugs out in the balancing series are 5,000, 1,000, 200, 20, 5, 2, 1, the resistance of the coil lies between 62·27 and 62·28 ohms.

When plug 1 is in and the keys both down, let the deflection of the galvanometer light spot be 10 divisions to the left, and when plug 1 is out let it be 20 divisions to the right. Then the exact resistance is 62·2733, being one-third of a unit on the way to ·28, because 10 is one-third of 10 + 20.

Determine in this way the exact resistance of a series of coils of wire given to you at different temperatures. Obtain the required temperature by heating the coil in paraffin oil and keeping the oil *well* stirred. Enter up your observations in the annexed table. Plot a curve for each coil, showing its resistance at different temperatures. If R_0 is the resistance of any coil at zero Centigrade, and R its resistance at $t°$C., and a is its *temperature coefficient*, find a by inserting observed values of resistance in the equation

$$R = R_0 (1 + a\ t).$$

Note that a is the percentage change of resistance per degree Centigrade, and a may be expressed either in terms of the resistance at zero Centigrade or the resistance at any other temperature. Determine the mean value of a at 15°C. for each coil you are using. If R is the resistance of a coil at t°C. and R^1 its resistance at t^1C.,

since

$$R = R_0 (1 + a\ t)$$

and

$$R^1 = R_0 (1 + a\ t^1),$$

we have

$$\frac{R}{R^1} = \frac{1 + a\ t}{1 + a\ t^1}$$

or

$$a = \frac{R - R^1}{R^1 t - R t^1}$$

Obtain various values of a from pairs of observations taken at two different temperatures not far apart.

Measure also with this bridge the resistances of various circuits, some of which, like dynamo field magnet circuits, are highly inductive. In all cases note that the battery key should be pressed down *before* closing the galvanometer key.

THE DETERMINATION OF RESISTANCE WITH THE POST OFFICE PATTERN WHEATSTONE BRIDGE.

Observation No.	Ratio arms.		Balancing resistance. S =	Calculated resistance. R =	Temperature.	Nature of Coil.	Temperature coefficient a.
	P =	Q =					

THE DETERMINATION OF RESISTANCE WITH THE POST OFFICE PATTERN WHEATSTONE BRIDGE.

Observation No.	Ratio arms.		Balancing resistance. S =	Calculated resistance. R =	Temperature.	Nature of Coil.	Temperature coefficient a.
	Q =	P =					

ELECTRICAL LABORATORY NOTES AND FORMS.

No. 10.—ELEMENTARY.

Name......   *Date*.............

Ꮳhe Determination of Potential Difference by the Potentiometer.

The apparatus required for these experiments is a potentiometer, which may be either a simple divided wire potentiometer or a more complete form such as " Crompton's"; a sensitive suspended coil galvanometer, three large cells of a secondary battery, a couple of standard " Clark" cells, a variable resistance or rheostat, and some experimental cells to test.

The Student is recommended to sketch the arrangement of the apparatus and circuits.

The potentiometer in its simplest form consists of a fine wire $a\,b$ of platinoid or German silver (*see* diagram) stretched over a scale which is divided into 4,000 parts. This wire should have a resistance of 40 or 50 ohms. To the ends of this wire are attached three large secondary cells B, and these cells make a fall of potential down this wire of about six volts. One terminal of the galvanometer G is attached to one end of this slide wire a, and the other end of the galvanometer is attached to that pole of a Clark standard cell Ck, which is of the same sign as the pole of the secondary battery attached to the end of the slide wire to which the other galvanometer terminal is directly attached. The second terminal of the Clark cell is connected to a slider S moving over the slide wire, and by means of which a contact can be made at any point of the slide wire. The Clark cell is a standard of electromotive force, and if correctly set up has the following values for its electromotive force (E.M.F.) at various temperatures.

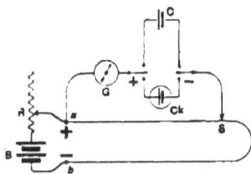

E.M.F. volts.	Temp. °C.	E.M.F. volts.	Temp. °C.	E.M.F. volts.	Temp. °C.
1·444	6	1·436	13	1·428	20
1·443	7	1·435	14	1·427	21
1·442	8	1·434	15	1·426	22
1·441	9	1·433	16	1·425	23
1·440	10	1·432	17	1·424	24
1·438	11	1·431	18	1·423	25
1·437	12	1·430	19		

The potentiometer is used to compare the potential difference or electromotive force of any other cell C with that of a Clark cell.

In order to avoid calculations, it is desirable to employ three cells of a secondary battery in connection with the slide wire, and to obtain a potential difference (P.D.) at the ends of the slide wire which is exactly equal to four volts. This is achieved as follows : A resistance R, which must be continuously variable, and the most convenient form of which is Mr. Shelford Bidwell's rheostat or Lord Kelvin's revolving rheostat, is inserted in series with the slide wire. The slider on the slide wire is set at that division on the scale which corresponds to the value of the Clark cell for the temperature of the day. Thus, if the temperature of the room is 18°C. the slider S is set to make contact at division 1,431, which corresponds to 1·431 volts. The resistance R is then varied until the galvanometer indicates no current when the slide key is down. When this is the case, we know that the fall of potential down the slide wire must be 1·431 volts for 1,431 divisions, and, therefore 4·000 volts for 4,000 divisions. This being done, any other cell or battery of which the E.M.F. is less than four volts, can be substituted for the Clark cell and connected up in the same way. If the position of the slider at which the galvanometer shows no current is ascertained, we then know the fall of potential down the slide wire due to the secondary cells is given directly in volts by the scale reading, and hence the E.M.F. of the cell or battery becomes known. Thus, assuming the resistance to have been set so that the Clark cell when at 18°C. balances at 1,431 on the slide wire and the other cell to be tested balances at 1,902 on the slide, this would indicate that the last cell has an E.M.F. of 1·902 volts.

Note that a high resistance must always be inserted in the circuit of the galvanometer, so that the Clark cell may not under any circumstances send any sensible current. The value of the Clark cell can only be assumed to remain constant if it is used merely as a standard of electromotive force. Take the precaution of checking the reading of the Clark cell continually during the progress of the experiments, and if the galvanometer gives any indication when the slider is made to touch at the scale reading corresponding to the Clark cell value when the Clark cell is connected on, then a small adjustment must be made of the variable resistance R to bring back the galvanometer to zero. It is not possible to get good readings unless the potentiometer, battery and galvanometer are well insulated by placing them on slips of ebonite or paraffined paper. Using the potentiometer in this way, determine the E.M.F. of a simple cell consisting of a carbon and a zinc plate placed in bichromate of potash solution, and determine the E.M.F. every five minutes for an hour or two. Determine also the E.M.F. at the terminals of various forms of single and two-fluid cells. Enter up your results in the Table on the opposite page.

The accuracy of the measurements will depend on the uniformity of the slide wire, since the assumption is made that equal lengths of wire correspond to equal falls in potential. The wire can be tested by measuring the length of wire which corresponds to a fall of potential equal to a known fraction of a volt, when a perfectly constant current is kept flowing through the wire. The wire should be rejected if non-uniform.

DETERMINATION OF POTENTIAL DIFFERENCE BY THE POTENTIOMETER.

Observation No.	Temperature of room.	E.M.F. of Clark cell.	Scale reading when Clark cell is in circuit.	Scale reading when E.M.F. to be determined is in circuit.	Value of unknown E.M.F. or P.D.	Time.	Remarks.
			-				

DETERMINATION OF POTENTIAL DIFFERENCE BY THE POTENTIOMETER.

Observation No.	Temperature of room.	E.M.F. of Clark cell.	Scale reading when Clark cell is in circuit.	Scale reading when E.M.F. to be determined is in circuit.	Value of unknown E.M.F. or P.D.	Time.	Remarks.

ELECTRICAL LABORATORY NOTES AND FORMS.

No. 11.—ELEMENTARY.

Name................   *Date*

The Measurement of a Current by the Potentiometer.

The apparatus required for these experiments is a potentiometer and the auxiliary apparatus as described in Elementary Form 10. In addition to this a series of low resistance standards of 1 ohm, 0·1 ohm and 0·01 ohm must be provided. These are best made of manganin strip to avoid temperature errors.

The Student is recommended to sketch the arrangement of the apparatus and circuits.

In Elementary Form 10 the use of the potentiometer has been explained as a means of determining a potential difference by comparison with the terminal potential difference or E.M.F. of a Clark standard cell. If a steady unidirectional current flows through a resistance, such resistance having sufficient sectional area and surface not to be sensibly heated by it, there is a fall of potential down this resistance, and this potential difference can be compared with the terminal potential difference of a Clark cell by means of the potentiometer. The condition of success is, however, that the total fall of volts down the resistance must be less than the fall which can be measured on the potentiometer, and which is generally less than two or four volts according as two or three secondary cells are used to work the potentiometer. Thus, if we have a resistance of 0·01 ohm, and we pass through this a current of 70 amperes, we have a fall of pressure of 0·7 volt down this resistance, or generally

$$\text{Drop in volts down resistance.} = \text{Value of resistance in ohms.} \times \text{Value of current in amperes.}$$

Hence, since 0·7 volt can be measured on the potentiometer, we have only to divide the value of the observed drop in volts by the value of the resistance to get the value of the current.

The accuracy with which the current can be measured is therefore dependent upon the accuracy of the low resistance. This resistance must be so constructed that it does not sensibly change its value when the largest current which it is intended to carry is sent through it. A resistance of this kind, say 0·1 ohm, is conveniently constructed by taking ten platinoid wires, each of No. 16 or No. 18 gauge and of such length as to measure one ohm. These ten wires, have their ends soldered to two large copper terminals. We have then a resistance of one-tenth of an ohm, capable of carrying without sensible heating ten amperes. If any current up to about ten amperes is sent through this resistance, and if *potential wires* are joined to the copper blocks and connected to the potentiometer, we can measure the fall in volts down this resistance. Then ten times the fall in volts is the value of the current in amperes passing through the resistance. In joining up the potential wires from the standard resistance to the potentiometer be careful to see that the right direction is given to the fall in potential. Thus, suppose the positive pole of the actuating secondary battery is joined to one end of the slide wire, we will call this A. Then to A must be joined, also through the galvanometer, the positive pole of the Clark cell and the potential wire which comes from the end of the resistance at which the current to be measured *enters*.

In making use of this method to standardize an amperemeter, ampere balance or galvanometer, the instrument to be standardized is joined in series (*see* diagram) with an appropriate resistance R. That is to say, a resistance such that when the largest current to be measured is sent through it the fall in volts down this resistance is not more, say, than two volts. Hence a series of such resistances must generally be used. The potential wires from the ends of this resistance are joined in the right direction to the potentiometer, and the potentiometer is set to read directly by the Clark cell Ck as described in Elementary Form No. 10. Various currents are then passed through the standard resistance and instrument to be standardized, and simultaneous readings taken by two observers of the scale reading of the instrument and the potentiometer reading, and the value of the scale reading of the instrument thus becomes known.

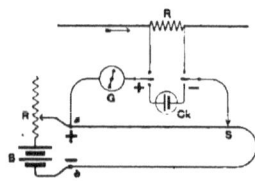

The current sent through the instrument to be standardized must be extremely steady. No good results can be obtained with a dynamo current. Nothing but the extremely steady current of a large secondary battery will give satisfactory results. In this case the instrument to be standardized must be joined up in series with an appropriate variable resistance to be able to vary the current sent through it and so obtain different scale readings.

Calibrate in this way one or more ammeters and enter up your results in the annexed form. Make a curve of errors for each ammeter, plotting the error + or − in terms of the scale reading.

MEASUREMENT OF A CURRENT BY THE POTENTIOMETER.

Observation. No.	Tempera- ture °C.	Slide wire reading for Clark cell.	Slide wire reading for fall of volts down resistance.	Standard resistance used.	Value of the current in amperes.	Reading of the ammeter.	Error of ammeter.

MEASUREMENT OF A CURRENT BY THE POTENTIOMETER.

Observation No.	Tempera- ture °C.	Slide wire reading for Clark cell.	Slide wire reading for fall of volts down resistance.	Standard resistance used.	Value of the current in amperes.	Reading of the ammeter.	Error of ammeter.

ELECTRICAL LABORATORY NOTES AND FORMS.

No. 12.—ELEMENTARY.

Name   *Date*

A Complete Report on a Primary Battery.

The apparatus required for these experiments is a potentiometer (Crompton's arrangement of the instrument being preferable), a sensitive suspended coil galvanometer, standard "Clark" cell, secondary cells for the potentiometer, and a series of resistances, 10, 1 and 0·1 ohm, of manganin or platinoid. The arrangement of the apparatus is the same as for testing voltage or current by the potentiometer.

The Student is recommended to sketch the arrangement of the apparatus and circuits.

The potentiometer is by far the most convenient and accurate method of making a complete examination of a primary or secondary cell. The cell to be tested (in this case a primary battery) is examined as follows :—

The zinc plate or active plate in the cell is first weighed carefully. If the zinc has been recently amalgamated it should before weighing be carefully wiped with some cotton wool or clean waste to remove any loosely adhering mercury. Its weight in grammes is then taken on a chemical balance and noted. The cell is then set up. Suppose, for example, it is a carbon-zinc cell with a single exciting fluid. The cell is set in action and a resistance of platinoid, r (*see* Diagram), say of one or two ohms, according to the current required, joined across the poles of the cell. The time of starting the current is noted. Two leading wires are joined to the terminals of the cell, c, and taken to the potentiometer, which is set up exactly as in the last exercise (*see* Elementary Forms Nos. 10 and 11) for measuring E.M.F. and current. The potential difference between the terminals of the cell to be tested, on *closed* circuit, is then measured and compared with that of a Clark cell, Ck (*see* Elementary Form No. 10), and the instant at which this measurement is made is noted. The platinoid resistance across the poles of the cell is next removed, and the potential difference between the terminals of the cell on *open* circuit is measured again, the time of making the measurement being again noted. Let the potential difference in volts or

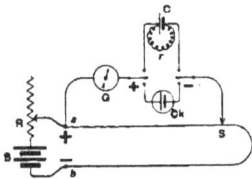

parts of a volt which is found between the terminals of the cell when it is on open circuit be denoted by V, and the potential difference which is found between the same points when the poles of the cell are short-circuited by the resistance of R ohms be called v. V will always be greater than v. If the internal resistance of the cell which exists at the instant of making the measurement v be called r, then by Ohm's law we have

$$\frac{v}{R} = \frac{V}{R + r}, \qquad \cdots \quad (1)$$

since the current which the cell is sending at that instant is equal to $\frac{v}{R}$. From the above equation we have $\qquad v\,R + v\,r = V\,R,$

or $\qquad\qquad\qquad\qquad r = \frac{(V - v)\,R}{v}.$ $\qquad\qquad$ (2)

This last equation gives us a value of the internal resistance of the cell at the instant when it is sending a current of $\frac{v}{r}$ amperes. These measurements of V and v must be repeated at intervals of time (the time being noted by the watch) for some hours, or until the cell has run down. The zinc plate can also be withdrawn at intervals, washed with clean water, dried and weighed. The results have then to be plotted out in a series of curves as follows:—Take a horizontal line on sectional paper to any suitable scale to represent time, and mark off intervals to represent hours and minutes. On this line erect at the appropriate places vertical lines to represent to some suitable scale

(*i.*) The open circuit volts V of the cell,

(*ii.*) The ampere current $\frac{v}{R}$ given by the cell,

(*iii.*) The internal resistance r of the cell calculated from equation (2).

Then draw curves through the tops of all these perpendiculars, and obtain curve of E.M.F., current and internal resistance. Integrate the area, by any means, included between the current curve and the base line, limited by the two extreme ordinates, and obtain the whole ampere-hours output of the cell up to the end of each hour. If the zinc has been weighed each hour, the loss in weight in grammes of the zinc can be obtained up to the end of each hour the experiment has lasted, and the ampere-hour output of the cell also obtained from this curve. Knowing that the full theoretical value of the consumption of zinc per ampere-hour of output should be 1·213 grammes per ampere-hour (since this is the electro-chemical equivalent of zinc), we can find the ratio between the actual ampere-hour output of the cell per gramme of zinc dissolved and the theoretical value up to and at the end of each hour. This gives us what is usually called the *efficiency* of the battery. It will be found on plotting the efficiency in terms of the time that it gradually falls in value as the time of experiment increases.

Taking the cell given to you, make in this way a full examination and report on it, and draw a complete chart of curves showing the variation with time of the E. M. F., current, and internal resistance of the cell and its efficiency.

TEST OF A PRIMARY CELL WITH THE POTENTIOMETER.

Temperature of Clark cell =   t =   °C.

Value of Clark cell E.M.F.   =   volts.

External resistance used to close
the circuit of cell tested = R   =   ohms.

Observation No.	Time.	Potential difference at terminals of cell when on open circuit = V volts.	Potential difference at terminals of cell when on closed circuit = r volts.	Calculated internal resistance of the cell = r, $r = \dfrac{V-v}{v}$ R.	Weight of zinc rod or plate w.	Efficiency of cell.

TEST OF A PRIMARY CELL WITH THE POTENTIOMETER.

Observation No.	Time.	Potential difference at terminals of cell when on open circuit $= V$ volts.	Potential difference at terminals of cell when on closed circuit $= v$ volts.	Calculated internal resistance of cell $= r$ $r = \dfrac{V - v}{v} R.$	Weight of zinc rod or plate $w.$	Efficiency of cell.

These Notes are copyright, and all rights of reproduction are reserved. They are arranged by Dr. J. A. FLEMING of University College, London, and are published by "The Electrician" Printing and Publishing Company, Limited, Salisbury Court, Fleet Street, London, England.

ELECTRICAL LABORATORY NOTES AND FORMS.

No. 13.—ELEMENTARY.

Name   *Date*

Standardization of a Voltmeter by the Potentiometer.

The apparatus required for the following experiments is a potentiometer, a standard "Clark" cell, and the galvanometer and secondary battery and rheostat for working the potentiometer as described in Elementary Form No. 10. In addition to these a divided resistance is required which must be capable of being placed safely and without sensible heating across the highest potential difference to be measured, and must be divisible into sections. Some voltmeters to check are also required.

The Student is recommended to sketch the arrangement of apparatus and circuits.

If a resistance composed of a great length of fine wire is placed across two points between which a constant difference of potential is maintained, there is a uniform fall of potential down this resistance. If this resistance is made in sections, the fall in potential down any section is to the fall in potential down the whole wire as the resistance of the section is to the resistance of the whole wire. In order to secure this uniformity the wire must be so arranged that all parts of the wire become equally heated when a current is passed through it. A divided resistance of this kind, made in such sections that one-hundredth or one-fiftieth of the drop in potential down the whole wire can be taken, is required in potentiometer tests of high voltage. A resistance divided in this way into sections of 100 : 1, 50 : 1, 20 : 1, or 10 : 1, is often called a volt box.

By means of a divided resistance of this kind, the potentiometer can be employed to compare high potentials such as 100 or 200 volts with the potential difference of the terminals of a Clark standard cell. In order to check a voltmeter by the potentiometer, we proceed as follows :—Let us assume that a voltmeter has to be checked of which the scale is divided to read from 40 to 100 volts, and it is desired to compare the indications of this instrument with a Clark standard cell as a standard of reference. The divided resistance, divided in the ratio of 100 : 1, is joined across the terminals of the voltmeter, and the two together are placed across the terminals of an incandescent lamp, or other resistance of such a character that 100 volts can be safely maintained across the terminals. In series with the lamp is placed a variable resistance, and by changing the value of this resistance the potential difference of the terminals of the lamp can be varied from 40 to 100 volts. A pair of potential wires are then taken from the small section of the divided resistance and joined to the potentiometer in the right direction (*see* Elementary Form No. 11). Let us suppose in the first case that the resistance in series with the lamp is all cut out, and that the potential difference (P.D.) at the terminals of the lamp is 100 volts. Then the P.D. at the terminals of the voltmeter to be calibrated is also 100 volts; the P.D. at the terminals of the divided resistance is also 100 volts, and the P.D. at the ends of the hundredth part of the resistance is *one* volt. This last P.D. can be accurately compared on the potentiometer with the P.D. of the terminals of a Clark standard cell. If the ratio of the divided resistance is 100 : 1, then the P.D. across the terminals of the lamp, and therefore across the terminals of the voltmeter, is 100 times that of the P.D. at the ends of the small section of the divided resistance. By varying the resistance in series with the lamp various voltages can be put upon the terminals of the voltmeter, and the actual observed scale readings of the instrument be compared with the values of the same potential difference as given by the potentiometer in terms of the Clark cell.

Suppose, for example, that the potentiometer is arranged as required for measuring potential difference, that there are 2,000 divisions on the slide wire, and that by secondary cells a potential difference of exactly 2 volts is maintained at the end of the slide wire, as determined in Elementary Form No. 10. Let the balance with the galvanometer be obtained when the potential wires from the ends of the small section of the divided resistance are connected to the beginning of the slide wire of the potentiometer and to a point on the slide wire indicated by 991 divisions on the scale. Then, if the potentiometer is properly adjusted, the true potential difference between the potential wires brought from the small section of the divided resistance is ·991 volt. If the divided resistance is divided in the ratio of 100 : 1, the P.D. between the ends of the whole resistance is 99·1 volts, and this is the true potential difference on the voltmeter terminals. If the scale reading of the voltmeter is, say, 102·5, this shows that the error of the voltmeter is + 3·4 volts, or that it reads 3·4 volts too high. By performing a similar measurement all along the scale, a scale error can be found and a curve of errors be constructed. Check in this way the voltmeters given to you, and construct for each a curve of errors by setting off distances on a horizontal line to represent the scale readings of the voltmeter and vertical lines drawn upwards (+) or downwards (−) to represent to an enlarged scale the error + or − of the instrument at these points. Note the vertical and horizontal scales need not be the same.

STANDARDIZATION OF A VOLTMETER BY THE POTENTIOMETER.

Ratio of sections of divided resistance used =

Observation No.	Temperature of Clark cell.	Value of E.M.F. of Clark cell.	Slide wire reading of potentiometer when Clark cell is connected on.	Slide wire reading of potentiometer when the section of divided resistance is connected on.	True P.D. at terminals of voltmeter calculated.	Observed scale reading of voltmeter.	Error of voltmeter.

STANDARDIZATION OF A VOLTMETER BY THE POTENTIOMETER.

Observation No.	Temperature of Clark cell.	Value of E.M.F. of Clark cell.	Slide wire reading of potentiometer when Clark cell is connected on.	Slide wire reading of potentiometer when the section of divided resistance is connected on.	True P.D. at terminals of voltmeter calculated.	Observed scale reading of voltmeter.	Error of voltmeter.

ELECTRICAL LABORATORY NOTES AND FORMS.

No. 14.—ELEMENTARY.

Name   *Date*

A Photometric Examination of an Incandescent Lamp.

The apparatus required for these experiments is a photometric gallery or bench, on which must be fixed (a) a standard of light, (b) a photometer, (c) an incandescent lamp to be examined. This last must have a suitable voltmeter across its terminals and an ammeter in series with it to read volts and current. The testing current should be supplied by secondary batteries. No satisfactory work can be done off dynamo machines directly. The means of standardizing the voltmeter and ammeter, and also of varying the working pressure, must be at hand.

The Student is recommended to sketch the arrangement of apparatus and circuits.

An incandescent lamp may have electric currents of various strengths, up to a certain limit, passed through it, and under these circumstances it gives varying candle-power and exhibits varying differences of potential (P.D.) between its terminals. If then we observe :—(a) The terminal *volts* V, or lamp P.D., (b) the *current* A in amperes through it, or lamp current, and (c) the candle-power, C.P., in any direction, we can derive

(*i.*)   The *hot resistance* R of the lamp in ohms, which is numerically equal to the quotient of V by A ;

(*ii.*)   The *total power* W in watts taken up in the lamp, since this is numerically equal to the product of A and V ;

(*iii.*)   The *watts per candle-power* w obtained by dividing W by C.P.

Hence if we measure and observe the quantities A, V, and C.P., we can calculate the quantities—

$$R = \frac{V}{A}, \qquad W = A\,V, \qquad w = \frac{W}{C.P.}$$

The lamp should be arranged as follows :—A lamp socket should have two long double flexible conductors (commonly called flexible cord) attached to its terminals. Two of these wires should be covered with red cotton and two with black, to avoid confusion. One pair (red and black) are attached to a convenient voltmeter having the necessary scale range, and which must previously have been calibrated by reference to an absolute standard (*see* Elementary Form No. 10). The other two wires (red and black) are connected to a circuit of constant potential, say a 100-volt circuit, passing through an ammeter, also calibrated, on one side, and a variable resistance on the other. The incandescent lamp to be tested is then placed in the socket, and the resistance adjusted until the ammeter shows a current not exceeding that which the lamp will bear. The voltmeter indicates then the P.D. on the terminals of the lamp.

This socket should be arranged to slide on the photometer bench on a suitable stand. At one end of the photometer bench is placed a suitable standard of light. The worst standard is the legal standard or parliamentary candle. In some cases a gas burner is used with a slit in front of the flame to adjust the light to a value of two candles, and this slit is called a " Methven " slit. This arrangement is by no means satisfactory, as variations of gas pressure, atmospheric pressure, atmospheric purity, and other causes create very marked variations in the illuminating power of the gas. The best arrangement is to employ a standard incandescent lamp the bulb of which is very much larger than is usually the case. Such a lamp, if not used much, will not blacken or deteriorate very quickly. This lamp is employed, in the first place, to standardise a number of similar secondary standard lamps, and these are used as the actual working standards on the photometer. The secondary standard lamps are marked with the mean candle-power which they give in a certain direction when a certain current is passed through them. The secondary standard lamp having been fixed on the photometer bench and adjusted by a separate ammeter and resistance to its normal candle-power, the other, or lamp to be tested, can be compared with it by using a photometer. This may be either a simple grease-spot disc, or any of the varieties of Ritchie wedge photometers, or more elaborate devices. Using the photometer, the Student has to practice the art of moving the standard lamp and lamp to be compared to such distances from the photometer that the two illuminated surfaces to be compared on the photometer are of equal brilliancy. This is comparatively easy when the lamps are in a similar state of incandescence, but it is much more difficult when one lamp is much brighter than the other. At the same instant that one observer takes the candle-power, two other observers must take the lamp volts and current. These observed values must be recorded in the proper columns in the Tables on the following pages 3 and 4. The resistance is then varied so as to alter the candle-power of the lamp, and a series of observations taken of the volts, current and candle-power of the lamp all the way up from dull initial incandescence to the highest volts the lamp will bear. The corresponding values of R, W and w are then calculated.

The candle-power should be observed in various directions, as follows :—

(*i.*) With the plane of the loop of filament parallel to the photometric disc,

(*ii.*) With the plane perpendicular to it,

(*iii.*) With the plane at an angle of 45° to it.

The mean value of these should be taken as the observed candle-power.

Having obtained all these observed and calculated quantities, they should be set out in curves on squared paper as follows :—

Set off vertical distances to represent to some suitable scale mean candle-power, and prepare a series of four curves in which horizontal distances are respectively current, volts, watts, and watts per candle-power. Also prepare a fifth curve in which horizontal distances are volts, and vertical distances are hot resistances.

Having obtained these curves, convert them into logarithmic curves by plotting horizontally and vertically the logarithms of amperes and candle-power, and the logarithms of volts and candle-power. These last curves should be nearly straight lines. Express the value of the candle-power, C.P., in terms of the amperes, A, and volts, V, in the form of two equations—

$$C.P. = P A^Q \qquad . \qquad . \qquad . \ . \ . \ (1)$$
$$C.P. = P' V^{Q'}, \qquad . \qquad . \ . \ . \ . \ . \ . \ (2)$$

where P, P', Q, Q' are some constants.

Since from (1) we have—
$$\log C.P. = \log P + Q \log A,$$

and from (2) $\qquad \log C.P. = \log P' + Q' \log V,$

it can be shown that the index Q or Q' is the tangent of the angle which the straight line obtained in plotting logarithms of amperes or volts and candle-power makes with the horizontal line, and the constants P and P' are the anti-logarithms of the intercepts on the vertical axis.

Obtain in this way equations for the candle-power of the lamp you are using, and compare the calculated and observed values of the candle-powers at various voltages.

A PHOTOMETRIC EXAMINATION OF AN INCANDESCENT LAMP.

Observation No.	Ampere currents A.	Terminal voltage V.	Mean candle-power C.P.	Watts A V = W.	Hot resistance $\frac{V}{A}$.	Watts per candle $\frac{A V}{W} = w$.	Name of lamp.

A PHOTOMETRIC EXAMINATION OF AN INCANDESCENT LAMP.

Observation No.	Ampere currents A.	Terminal voltage V.	Mean candle-power C.P.	Watts A V = W.	Hot resistance $\dfrac{V}{A}$.	Watts per candle $\dfrac{A V}{W} = \kappa$.	Name of lamp.

ELECTRICAL LABORATORY NOTES AND FORMS.

No. 15.—ELEMENTARY.

Name ....   *Date*

The Determination of the Absorptive Powers of Semi=Transparent Screens.

The apparatus required for these experiments is a photometric bench provided with a photometer, standard of light, incandescent lamps, and instruments for measuring currents and potential differences. Also various semi-opaque and translucent shades, glasses and screens.

The Student is recommended to sketch the arrangement of the apparatus.

In the practical use of electric lights, whether arc or incandescent, it is usual to employ various forms of semi-opaque or translucent globes, shades or screens to diffuse the light. The object of using these shades is to obtain a larger and less dazzling illuminating surface. These screens, however, cut off a considerable percentage of light. It is required to determine for each particular shade or globe what this percentage is. An incandescent lamp is placed on the photometer bench and adjusted carefully to its normal volts. The candle-power of the lamp is then carefully taken in various positions. If the lamp is placed vertically, and the vertical line through the centre of the lamp is called the *axis*, the candle-power should be taken in the direction of the axis, and in three directions horizontally, one with the plane of the filament parallel to the photometric disc, one with plane perpendicular, and the other with the plane at 45°. The current through the lamp and the terminal volts should be kept constant, and be taken at the same time. When this is done, a shade, or globe, or screen is put over the lamp, which shade may be any of the ground glass, semi-opaque, or fancy shades sold for the purpose, and the candle-power taken at the same volts and current again. The difference of the candle-power in any direction with the globe on, and the candle-power in that direction with the globe off, gives the loss in candle-power due to the globe or shade, and this may be expressed as a percentage of the original candle-power in that direction. Using various screens, such as ground glass, semi-opaque glass, tissue paper, tracing cloth, writing paper, frosted glass, determine the percentage loss of light due to these screens when placed over a 16-candle-power incandescent lamp, and record the results in the Tables provided on pages 3 and 4.

Note the following precautions in using the photometer :—The eye is not in a suitable condition for discriminating small difference of brightness of two surfaces if it has been recently stimulated with bright light. It is therefore impossible to make good photometer measurements unless the eyes have been preserved in dim light for some time. Photometer measurements should always be made, therefore, in a large well-ventilated photometric room or gallery, the walls of which are painted dead black, and in which no lights are used except those being compared. The observers must keep in this room some time before making the measurements. The eyes of different observers differ greatly in sensibility. Each observation should therefore be taken several times by different observers and the names of the observers recorded.

It is impossible to get good results when candles or gas flames are used as standards of light in badly ventilated photometric rooms, as the supply of oxygen for the flame is insufficient. In making a comparison of the illuminating power of one light with that of a standard, when once the approximate distances have been discovered at which the lights balance on the photometer, the measured light or the photometric disc should be oscillated to and fro in gradually diminishing arcs until an exact balance is found. It is found that this process assists the eye in discriminating between a small difference in *brightness* of the illuminated surfaces compared and any small difference in *colour* which may exist. The student should therefore practise comparing one incandescent lamp at normal brightness with one of which the filament is kept at a red heat, and endeavour to overcome the difficulties attending the photometry of lights of different colour.

THE DETERMINATION OF THE ABSORPTIVE POWERS OF SEMI-TRANSPARENT SCREENS.

Observation No.	Distance of standard lamp from photometric disc.	Illuminating power of the standard lamp.	Distance of in-candescent lamp from photometric disc.	Illuminating power of the incandescent lamp, *uncovered.*	Illuminating power of the incandescent lamp, *covered with screen.*	Nature of screen.	Percentage absorption of screen.

THE DETERMINATION OF THE ABSORPTIVE POWERS OF SEMI-TRANSPARENT SCREENS.

Observation No.	Distance of standard lamp from photometric disc	Illuminating power of the standard lamp.	Distance of in- candescent lamp from photometric disc.	Illuminating power of the incandescen lamp, uncovered.	Illuminating power of the incandescent lamp, covered with screen.	Nature of screen.	Percentage absorption of screen.

ELECTRICAL LABORATORY NOTES AND FORMS.

No. 16.—ELEMENTARY.

Name   *Date*

The Determination of the Reflective Power of Various Surfaces.

The apparatus required for these experiments is a photometric bench fitted up as described in Elementary Form No. 14 for the photometric measurement of incandescent lamps, and provided with a standard of light. The actual means of comparison, whether Bunsen grease-spot disc, wedge, or other photometer, must be enclosed in a box provided with two small round apertures on either side, and be so placed that rays from the incandescent lamp being measured can only fall on the disc when sent along the direction of the photometer bench. The walls of the photometer room or gallery should be painted dead black, or be lined with black velveteen.

The Student is recommended to sketch the arrangement of apparatus and circuits.

When light from any source falls on a polished or reflecting surface, some of the light is reflected regularly according to the laws of reflection. The percentage value, expressing the ratio of the intensity of the reflected ray to the incident ray, is called the *coefficient of reflection*. Some of the light is also irregularly reflected or scattered. The value of the coefficient of reflection depends upon the angle of incidence of the ray. Nearly all surfaces reflect a much larger proportion of the incident light at large angles of incidence than at small angles.

In order to determine the coefficient of reflection of a plane mirror or looking-glass at various angles of incidence the following procedure is adopted:—A standard incandescent lamp is placed as a standard of light on the photometer bench, and means are taken, by the employment of a resistance, to keep the current through it perfectly constant during the experiments. Another incandescent lamp is placed on the other side of the photometer disc, and in the same way its illuminating power is preserved constant by keeping a constant and regulated current passing through it. The most convenient way of doing this is by means of a carbon resistance, which consists of plates of carbon more or less compressed with a screw. Thus can very fine adjustments of resistance be made, and the current in a circuit be governed with great exactness.

The second incandescent lamp being placed on the photometer bench is first photometered directly against the standard in some definite position; say, with the plane of the looped filament parallel to the plane of the photometric disc.

A glass mirror or looking-glass, or silvered mirrors of about 6 inches by 4 inches in size, is then placed on the photometer bench, with its plane at any angle to the axis of the bench, and the incandescent lamp is moved into such a position that its light cannot fall on the photometric disc, except after reflection at this angle in the mirror. The incandescent lamp is moved to such a distance that the balance of illumination is obtained on the photometric disc, between the light of the standard lamp and that of the incandescent lamp *after* reflection at the desired angle in the mirror. The distance of the incandescent lamp from the photometer disc is then measured.

The illuminating power of the incandescent lamp is to that of the standard lamp as the square of the distance of the incandescent lamp from the disc is to the square of the distance of the standard from the disc. The coefficient of reflection of the mirror at the determined angle is then obtained by expressing as a percentage the ratio of the candle-power of the incandescent lamp, measured after reflection in the mirror to the value measured directly without reflection. Thus, if the value of the standard lamp is 16 candles, and the incandescent lamp is found also to have an illuminating power of 16 candles when measured directly, but only of 14 candle-power when the ray is measured after reflection at an angle of 45° from a plane mirror, the coefficient of reflection is $\frac{14}{16}$ = 87 per cent. In the same way the coefficient of reflection may be determined at various angles and for various surfaces. The chief precaution which must be taken is to test whether any light from the incandescent lamp when in the second position can fall on the photometric disc when the mirror or reflecting surface is taken away. If this is found to be the case, scattered light is getting into the photometer box, and this must be prevented by screens of black velvet suitably placed.

Measure in this way at various angles of incidence the coefficient of reflection of a plane mirror or looking glass, of polished metal plates, white paper, brown paper, and painted wood, and enter up the results in the annexed tables.

REFLECTIVE POWER OF VARIOUS SURFACES.

Observation No.	Candle-power of standard lamp $= I$.	Distance of standard from the photometric disc $= d_1$.	Distance of incandescent lamp from the photometric disc $= d_2$.	Illuminating power or candle-power of incandescent lamp measured directly $= I_1 = I \left(\dfrac{d_2}{d_1}\right)^2$.	Illuminating power or candle-power of incandescent lamp measured after reflection $= I_2$.	Angle of reflection $= \theta$.	Coefficient of reflection $= \dfrac{I_2}{I_1}$.

REFLECTIVE POWER OF VARIOUS SURFACES.

Observation No.	Candle-power of standard lamp $= I.$	Distance of standard from the photometric disc $= d_1.$	Distance of incandescent lamp from the photometric disc $= d_2.$	Illuminating power or candle-power of incandescent lamp measured directly $= I_1 = I \left(\dfrac{d_2}{d_1} \right)^2.$	Illuminating power or candle-power of incandescent lamp measured after reflection $= I_2.$	Angle of reflection $= \theta.$	Coefficient of reflection $= \dfrac{I_2}{I_1}.$

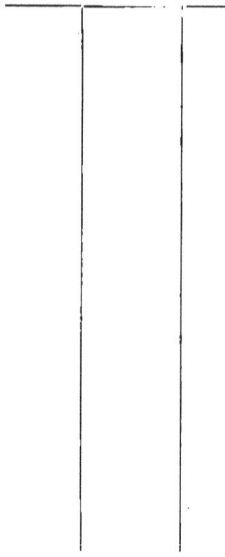

ELECTRICAL LABORATORY NOTES AND FORMS.

Name   *Date*

The Determination of the Electrical Efficiency of an Electromotor by the Cradle Method.

The apparatus required for this experiment is a small electromotor, which may be a shunt or series motor. The motor is balanced on a "Brackett" cradle. A load is put upon the motor by a cord wound round its pulley. An ammeter, voltmeter and resistances must be provided with which to measure and regulate the electrical power supplied to the motor.

The Student is recommended to sketch the arrangement and note the details of the cradle.

The *efficiency* of any transforming device such as an electromotor is a term employed to denote the ratio, generally expressed as a percentage, between the power given out and the power absorbed by the device, both powers being measured in the same kind of units. In the case of a continuous current electromotor an electric current A is put into the motor, and a certain difference of potential, V, exists between the terminals of the motor. The product A V, A being measured in amperes and V measured in volts, is the measure of the power in watts put into the motor. The motor converts part of this electrical power into mechanical power.

If a heavy body rests upon a plane surface there is a friction between the two surfaces which necessitates the exertion of a force to move the body along the surface. Let F be the average force measured in any units required to move the body a distance, D. Then the product F D represents in certain units the *work* done in moving the body. Apply this to the case of a strap or rope taken round a pulley. Let the belt be passed round one-half of the pulley, and let the sides of the belt be brought parallel to each other, and be kept stretched with a tension T. Next let the pulley be forcibly turned round under the belt. The friction between the belt and pulley will tighten one side of the belt, or increase its tension, say to a value T_1, and the tension of the other side will be decreased to a value, say T_2. If the pulley, having a radius R, is turned once round against this friction, the work done is equivalent to moving a body through a distance $2\pi R$ against a force $T_1 - T_2$. Hence the mechanical work done is $2\pi R\,(T_1 - T_2)$. If the pulley is turned round N times a minute, the work done per second at the rate of doing work is then equal to

$$\frac{2\pi N R (T_1 - T_2)}{60}.$$

In this last expresssion, $\dfrac{2\pi\,N}{60}$, which may be called ω, is the *angular velocity* of the pulley, and the other part of the expression, viz., $R\,(T_1 - T_2)$, is the *couple* or *torque* exerted in turning the pulley round. Call this F. Then if w is the work done per second, or the power applied to effect this rotation, we have

$$w = \omega\, F. \qquad (1)$$

Suppose, then, that a continuous current electromotor has a current of A amperes put into it and a terminal potential difference (P.D.) of V volts. The power, W, applied to the motor electrically is A V watts. If we measure the angular velocity ω of the moving part of the motor, and if we can measure the couple or torque F which it exerts when running at any required speed, we see by the above that the work done per second by the motor, or the power given out by it, w is equal to $\omega\,$F. Hence the efficiency e of the motor is given by

$$e = \frac{\omega\,F}{A\,V} = \frac{w}{W}. \qquad (2)$$

The most simple method of determining the torque or couple F in a running motor is by means of a "Brackett's" cradle. The motor is fixed on a small wooden cradle balanced on knife edges, and of such size that the line through the knife edges passes through the axis of the motor. To this cradle is attached a long arm and sliding weight. The motor must be screwed to the cradle, and by means of counterpoise weights the centre of gravity of the mass must be brought up or down it so as to be on the axis line through the knife edges. The cradle is then balanced with its knife edges on steel planes carried on a suitable frame. A cord is then wrapped once round the motor pulley and stretched at both ends, so that, when pulled in opposite directions, friction is put upon the pulley when current is put into the motor, and this cord brake is applied to make the motor give out work. The armature of the motor, by its reaction on the fields, tends to tilt the fields, and therefore the whole motor, on its cradle in an opposite direction. The motor can be restored to its initial position by sliding a weight on the balance arm. When this is done, the product of this weight P and the distance a at which it is placed from the axis gives us the couple P a required to keep the field magnets from turning round, and therefore represents the couple which the armature is exerting. Hence P a, so measured in feet and pounds, or centimetres and grammes, gives us the value of the motor torque or couple F. The speed of the motor is at the same time observed, and also the value of the current A and the terminal volts V. Carry out in this manner a series of experiments with a small motor, and determine its efficiency at various speeds and loads, from the equation (2) above.

NOTE.—1 horse power = 746 watts = 550 foot pounds per second. Hence, if the torque F is measured in feet-pounds, $\omega\,$F must be reduced to watts by proper multiplication by 746 and division by 550.

EFFICIENCY OF A SMALL MOTOR BY THE CRADLE METHOD.

Observation No.	Speed N.	Current in amperes = A.	Terminal P.D. in volts = V.	Balance weight in lbs. = P.	Length of arm in feet = a.	Power put in in watts = W.	Power taken out in watts = w.	Efficiency of motor = e.

Observation No.	Speed N.	Current in amperes = A.	Terminal P.D. in volts = V.	Balance weight in lbs. = P.	Length of arm in feet = a.	Power put in in watts. = W.	Power taken out in watts = w.	Efficiency of motor = e.

ELECTRICAL LABORATORY NOTES AND FORMS.

No. 18.—ELEMENTARY.

Name   *Date*

The Determination of the Efficiency of an Electromotor by the Brake Method.

The apparatus required for these experiments is a half or one horse-power shunt electromotor, an ammeter, and a voltmeter for taking current and volts, a speed counter, resistances for regulating the current, and a rope brake. This latter consists of a couple of half-inch diameter ropes of sufficient length to pass well round the motor flywheel, and having a spring balance at one end and a weight at the other.

The Student is recommended to sketch the arrangement of the brake and weight.

In the previous Form (No. 17, Elementary) the theory of the efficiency determination of an electromotor has been explained. In the case of large single motors which cannot easily be moved from their positions, and in which the cradle method is not convenient to apply, resource must be had to a "brake method" of determining the efficiency. The motor to be tested should be provided with a flywheel. Assuming the motor to be of about half or one horse-power, or more, this flywheel may be 10in. or 12in. in diameter, and 1in. or 2in. in thickness. It may be a simple cast-iron disc turned up on the edge. Two ropes are provided, of about half an inch in diameter and about three feet in length. These are laid parallel to each other along the edge of the flywheel, and are kept from slipping off the edge of the wheel by three little cleats of wood or metal which are fastened to the outside of the ropes, and have flanges which just prevent them from slipping sideways off the wheel. One end of this rope has a weight attached to it, and the other end carries a spring balance, which should be stretched to about half its full extent by the weight attached to the other end of the rope. This rope brake is laid over the pulley flywheel so that the weight hangs down on one side, and the spring balance on the other. The free end of the spring balance is attached to a hook screwed into the floor. When the motor is at rest the weight will stretch the rope, and the reading on the spring balance will be about equal to the value of the weight. The tension of one side of the rope brake is equal to the numerical value of the weight attached to that side. Call it W lb. The tension of the other side of the rope brake is given by the reading of the spring balance. Suppose it is W¹ lb. The difference of tensions of the two parallel sides of the rope brake is then W¹ − W lb.

If, now, the motor is set in motion by sending into it a current of A amperes, and if the potential difference at the terminals of the motor is V volts, then the power P supplied electrically to the motor is A V watts. If R is the radius of the pulley flywheel, and if t is the thickness of the rope, the couple or torque against which the motor does work is $\left(R + \frac{t}{2}\right)(W^1 - W) = F$, and the rate at which work is done by the motor, or the power in watts given out by it on the flywheel, is equal to

$$\frac{746}{550} \times 2\pi \frac{N}{60}\left(R + \frac{t}{2}\right)(W^1 - W) = p,$$

where N is the number of revolutions of the motor per minute.

Note that if N is measured in revolutions per minute, R and t in feet, or fractions of a foot, and W^1 and W in pounds, the value of p is given in foot-pounds per second. Hence, this must be *divided* by 550 to reduce to *horse-power* and *multiplied* by 746 to reduce again to watts, since 550 foot-pounds per second is one horse-power, and one horse-power is 746 watts. The ratio of p to P, expressed as a percentage, is the efficiency e of the motor. In performing the experiment several observers are needed. One to take the speed of the motor by counting with a counter the revolutions per minute of the shaft; another to read the ammeter and voltmeter, which are put respectively in the motor circuit and across the main terminals; and a third to attend to the brake, and read the spring balance. The load on the motor is varied by varying the weight on the one end of the rope brake.

Provided with these appliances, the observers should take a series of measurements of the efficiency of the motor at the same speed but various loads, from the lowest possible load up to the full load the motor will safely carry. The results should be entered up in the appended tables, and then plotted out in the form of an efficiency curve, as follows :—A horizontal line is taken, on which the values of the power output of the motor are set off, best represented in fractions of full load taken as unity. Vertical ordinates are then taken to represent the efficiency e of the motor at these various fractions of full load, and an efficiency curve drawn showing the efficiency for any speed. A series of efficiency curves may be drawn corresponding to different speeds. The efficiency is the ratio of p to P expressed as a percentage, or is the numerical value of the power given out in watts when the power taken in or absorbed is 100 watts. The power taken in may be partly employed to excite the fields, in which case the value of this should be determined by measuring the field current and the voltage at the terminals of the field.

DETERMINATION OF THE EFFICIENCY OF AN ELECTRO-MOTOR BY THE BRAKE METHOD.

R = Radius of pulley flywheel =   t = Thickness of rope brake =

Observation No.	Revolutions per minute = N.	Ampere current taken in = A.	Terminal voltage of motor = V.	Weight on rope brake = W.	Spring balance reading = W¹.	Efficiency = e.	Fraction of full load carried.

DETERMINATION OF THE EFFICIENCY OF AN ELECTRO-MOTOR BY THE BRAKE METHOD.

Observation No.	Revolutions per minute = N.	Ampere current taken in = A.	Terminal voltage of motor = V.	Weight on rope brake = W.	Spring balance reading = W'.	Efficiency = e.	Fraction of full load carried.

ELECTRICAL LABORATORY NOTES AND FORMS.

Name   *Date*

The Efficiency Test of a Combined Motor-Generator Plant.

The plant required for these tests is a pair of dynamos with shafts coupled together, and which are preferably bolted down on the same bedplate. One of these machines should be a continuous current motor, suitable for being worked off the circuits of the laboratory, and the other may be either an alternator or a continuous current dynamo. Ammeters and voltmeters must be provided to measure the ingoing and outgoing currents and powers, and also a resistance of capacity enough to take up the whole of the power given out by the dynamo.

The Student is recommended to sketch the arrangement of the circuits.

When any electrical transforming device, such as a motor or lamp, has a continuous electric current passed through it, this current experiences a fall in pressure. The product of the current measured in amperes (let its value be A), and the fall in pressure measured in volts (let its value be V), gives us the power W taken up in the device measured in watts, so that

$$A V = W.$$

This power, in the case of a motor, is used up in several ways. (i.) Part of it is employed to excite the field magnets. If a is the ampere value of the current in the fields, and v is the potential difference between the ends of the field circuit, then $a v$ is the power in watts used in excitation. Another part of the power is used up in heating the armature circuits. If a^1 is the current through the armature when the machine is running, and if r is the resistance of the armature when warm and running, then $a^2 r$ is the power wasted in the armature. A third part of the power is wasted in making *eddy currents* of electricity in the iron core, and in the circuits. Call this d. A fourth part of the power is employed in overcoming the friction at the bearings of the machine, and this may be called f. Lastly, a certain part of the power applied is yielded up in the form of mechanical power on the shaft. If the motor exerts a *torque* or couple T, and if it runs with an angular velocity ω ($\omega = 2\pi n$, where n = revolutions per second), the external rate of doing work is ωT. Hence, adding all these modes of consumption of the external power applied together, we have

$$W = A V = a v + a^2 r + d + f + \omega T.$$

The *efficiency* of the motor is the ratio between the useful mechanical power which comes out, viz., ω T, and the whole electrical power put in, viz., A V. Hence, if E stands for the efficiency of the motor,

$$E = \frac{\omega\, T}{A\, V}.$$

Suppose, in the next place, that this motor is employed to run a generator, either an alternator or a *continuator*—*i.e.*, a continuous current dynamo—the power applied to the shaft of the continuator is the same as that given out by the motor, and this generator transforms part of this mechanical power back into electrical power. Let us suppose the alternator or continuator gives out a current of A_1 amperes, and that, when this current is taken up in a suitable resistance of wire, the potential difference between the terminals of the generator is V_1, then the power given out by the generator is $A_1 V_1$ watts $= W_1$, and the power put into the motor is $A V = W$. Hence, the combined efficiency of the motor-generator (call it e) is

$$e = \frac{A_1 V_1}{A\, V} = \frac{W_1}{W}.$$

With the motor-generator at your disposal, carry out a series of tests, measuring the ingoing and outcoming current and volts at various loads, and obtain the efficiency of the combination at these loads. Enter your observations in the annexed tables.

Plot down your results in a curve in the following manner :—Take a horizontal line to represent the outgoing power of the combination, and set off horizontal distances to represent the value of the outgoing watts $W_1 = A_1 V_1$. Then erect perpendiculars to represent to any suitable scale the efficiency e of the combination, and draw an efficiency curve.

Note that, as the load on the machines is varied, it may be necessary to vary the field exciting current of the motor in order to keep the speed of the combination constant. For this purpose the motor should be a *shunt*-wound motor, and resistance should be provided to insert in the field circuit of this motor. The constancy of the speed should be ensured by having a speed indicator or velocity meter attached to the shaft. The coupling of the shafts of the motor and generator should be done directly by some form of flexible coupling. It is not satisfactory to connect them by a belt, because the slip of this belt would introduce errors. If the motor and generator are two similar continuous current machines with pulleys on their shafts, they can easily be coupled as follows :—Bed down the machines so that their pulleys are in opposition and shafts in one straight line, and let the pulleys overhang the shafts. In the flange of each pulley drill four or six small holes. Take a broad piece of leather, sufficiently long to go once round the pulleys and sufficiently wide to extend over half of each pulley, fasten this leather ring to each pulley by small bolts and nuts, going through the holes in the pulley flanges and the leather. The machines will then have their shafts coupled by a flexible coupling.

THE EFFICIENCY TEST OF A MOTOR-GENERATOR.

Observation No.	Current into motor = A.	P.D. at motor terminals = V.	Current out of generator = A₁.	P.D. at terminals of generator = V₁.	Efficiency of motor generator = e.	Speed of combination N.

THE EFFICIENCY TEST OF A MOTOR-GENERATOR.

Observation No.	Current into motor = A.	P.D. at motor terminals = V.	Current out of generator = A₁.	P.D. at terminals of generator = V₁.	Efficiency of motor generator = e.	Speed of combination N.

ELECTRICAL LABORATORY NOTES AND FORMS.

No. 20.—ELEMENTARY.

Name............................   Date........

Test of a Gas Engine and Dynamo Plant.

The apparatus required for this test is a gas engine and dynamo. The gas engine must be provided with an indicator and indicating gear, and with a counter for taking the number of explosions and the speed. It is desirable also to measure the water passing through the cylinder jacket, and the temperature of the incoming and outgoing water. The electrical power given out by the dynamo should be taken up in resistances. In default of anything better, an iron wire resistance may be used, but a bare platinoid wire immersed in running water forms a better absorber of power.

The Students assisting at this test should all make sketches and diagrams of the circuits and connections.

In taking a complete test of a motive-power or electrical power plant of any kind, we wish to discover all the ways in which the energy supplied to the plant is transformed and used, and the relative amounts of the same. For this purpose there is no better plan than to make a kind of balance sheet of energy in which we put on the debtor side all the energy given to the plant, and on the creditor side all the items of energy given out by it.

It will be assumed that the gas engine used in this test is the ordinary double-cycle gas engine, and that the dynamo is the ordinary shunt wound dynamo as used for accumulator charging, since this plant is most frequently found in electrical laboratories. The quantities which have to be measured and recorded are as follows (one observer should be delegated specially to record each item during the run):—

1. The volume of gas used by the engine = G. This must be recorded by a meter on the intake gas pipe of the engine. If there is no special meter here the test should be made at a time when all other gas lights in the building can be put out and the record taken on the main meter, but this is not so satisfactory as a special meter.

2. The amount of cooling water used in the cylinder jacket = W. Whenever possible it is desirable to have a water meter put on a direct supply of water to the jacket and to take the temperature of the incoming and outgoing water. This, combined with the known volume of water used, will give the thermal units removed from the cylinder during the trial.

3. If possible the oil used during the trial should be noted. See that all bearings and lubricators are in good order and filled up before starting.

4. The number of revolutions of the engine per minute = N must be taken.

5. The number of explosions per minute = n must be counted. This can be taken by an automatic counter, which is moved by the inlet gas valve lever at each admission of gas to the cylinder.

6. The indicator diagrams must be taken. The ordinary Richards High-Speed Indicator may be used, or else the newer form by Wayne. Cards should be taken at regular intervals during the run, and the I.H.P. worked out at once.

7. The ampere-current = A given out by the dynamo must be noted at intervals, the time being taken when the observation is made.

8. The volts = V at the terminals of the dynamo are to be recorded at the same instant that the current is measured. The product of A and V gives the power in watts being given out by the dynamo.

The mean value of A V during the run gives the mean power in watts so generated. The quotient of $\dfrac{A\,V}{746}$ gives the same in horse-power, and the product of this last and the time of the run in hours gives us the energy in horse-power hours given out by the dynamo in the run.

One Board of Trade unit = 1,000 watt-hours = $\dfrac{1,000}{746}$ (= 1·3 nearly) horse-power hours of energy.

The above eight quantities should be simultaneously observed during a run of two or three hours, and each observer should afterwards get from the others all the observed values and enter them up on the Tables appended. It is a good plan to appoint one observer as fugleman, who gives the word when the observation is to be taken, and notes by a watch the instant when it is so done. In this way a regular series of observations at intervals of about 10 minutes or so can be obtained, and the readings of all the instruments are simultaneous.

The observations are then to be reduced as follows : From the indicator diagrams the mean pressure = P during the explosion must be obtained. This can be done best by the planimeter (Amsler's), taking the whole area and dividing it by the length of the diagram. Knowing the spring and scale, we obtain the mean pressure in pounds per square inch. The piston area = a and stroke = s must be measured in feet and square inches, and the indicated horse-power = H.P. calculated from the formula

$$\text{I.H.P.} = \frac{a\,P\,s\,n}{33,000},$$

where n is the number of explosions per minute.

We have then a record of the I.H.P. at various intervals of time. If these are equidistant we can obtain easily, by taking the mean of them, the mean I.H.P. during the run. From the observed volume of gas = G used we can obtain the cubic feet of gas per I.H.P., and from the water used the gallons of cooling water circulated per I.H.P.

The mean electrical output of the dynamo in horse-power is also obtained by multiplying the mean value of the product A V and 746. This is sometimes called the electrical horse-power of the dynamo, and is denoted by E.H.P., just as the indicated horse-power of the engine is denoted by I.H.P. The ratio of E.H.P. to I.H.P. is called the *combined efficiency* of the engine and dynamo. The gas and water used can also be reckoned out in terms of the electrical horse-power.

The following are the data of the engine and dynamo used in this trial :—

GAS ENGINE No.			DYNAMO No. . __		
Makers, Messrs.	. _ _____ __		Makers, Messrs.	_____	
Stroke	= s =	inches.	Normal speed	=	revs. per min.
Piston area	= a =	square inches.	Normal volts	=	volts.
Normal speed	=	revs. per min.	Full current	=	amperes.
Type of engine	=		Type of dynamo =		

Resistance of fields

Resistance of armature

Current in fields

GAS ENGINE AND DYNAMO TRIAL.

Made at.. *Date*..

TABLE I.

Time of observation.	Current from dynamo = A.	Volts at dynamo terminals = V.	Number of explosions in one minute = n.	Gas meter reading = G.	Water meter reading = w.	I.H.P. from card taken during the minute.	Efficiency E.H.P. ÷ I.H.P.

GAS ENGINE AND DYNAMO TRIAL.

Made at...   *Date*

TABLE I.

Time of observation.	Current from dynamo = A.	Volts at dynamo terminals = V.	Number of explosions in one minute = n.	Gas meter reading = G.	Water meter reading = w.	I.H.P. from card taken during the minute.	Efficiency = E.H.P./I.H.P.

TABLE II.

Mean I.H.P. during run.	Mean E.H.P. during run.	Total gas used during run.	Total water circulated during run.	Cubic feet of gas per mean I.H.P.	Gallons of water per mean I.H.P.	Cubic feet of gas per mean E.H.P.	Gallons of water circulated per mean E.H.P.

ELECTRICAL LABORATORY NOTES AND FORMS.

No. 21.—ADVANCED (No. 1).

Name   Date

Determination of the Specific Electrical Resistance
of a Sample of Metallic Wire.

The apparatus required for this determination is a Wheatstone's Bridge set, coupled with a sensitive movable coil galvanometer. The coils of the bridge should previously have been tested and compared with a standard coil or coils, of which the true electrical resistance is known at definite temperatures. For the measurement of specific gravity a good chemical balance and weights are required. For the measurement of length a metre metal scale and beam compass are needful.

The Student is recommended to make sketches of the arrangement of the apparatus, and to draw carefully on squared paper the curves representing the observation.

The specific electrical resistance of any material is the resistance of a mass of known volume or weight of it in a certain form and taken between defined surfaces. By *volume specific resistance* is meant the electrical resistance of one cubic centimetre of the material measured between opposed surfaces of the cube and expressed in absolute C.G.S. units. By *mass specific resistance* is meant the resistance of a known mass of the material, viz., one gramme, in the form of a circular-sectioned wire one metre in length, the resistance being taken between the ends of this wire.

If the resistance of a uniform circular-sectioned wire of length l centimetres and diameter d centimetres is equal to R ohms, then, if ρ is the volume specific resistance as above defined, we have—

$$R = \frac{4\,\rho\,l}{10^9\,\pi\,d^2},$$

or

$$\rho = \frac{10^9\,\pi\,d^2\,R}{4\,l}\,. \qquad (1)$$

The multiplier 10^9 is required to reduce resistance measured in ohms to its value in C.G.S. absolute units.

If w is the weight in grammes of this wire and s is its specific gravity, then

$$\frac{\pi\,d^2}{4}\,l\,s = w,\ \text{or}\ \pi\,d^2 = \frac{4\,w}{l\,s}\,. \qquad (2)$$

Hence, by substitution from (2) or (1) we get—

$$\rho = \frac{10^9\,R\,w}{l^2\,s}, \qquad (3)$$

and ρ thus becomes known when we know the resistance in ohms of a length l centimetres of the wire, its weight w in grammes, and its specific gravity s. The determination of the volume specific resistance requires, therefore, the determination of the length, weight, specific gravity and electrical resistance of a piece of circular-sectioned wire drawn from the material. In the case of most ordinary metals, the most convenient diameter for the wire to be used is about 0·02 of an inch, or 0·5 millimetre, and the length may be from one to six metres.

Let us assume that the volume specific resistance of a sample of copper wire is to be determined. The wire having been carefully drawn down to about the above diameter, the operations are conducted in the following order : On a stout plank of deal small brass plates are screwed, with centres at distances of one metre apart. Transverse scratches are made on these plates by means of beam compasses, such scratches being exactly one metre apart. Two lengths of the wire to be measured are then laid over the board and carefully straightened, and are each cut two centimetres longer than three metres. These two wires must then be covered with cotton, or silk wound on, to insulate them. A terminal rod of high conductivity copper of about 0·1 inch diameter and 1 foot in length is then carefully soldered to one end of each of the two lengths of insulated wire, the overlap of wire against the rod being exactly one centimetre. The other ends of the two lengths are carefully twisted together; one centimetre of the length, exactly, being employed for the twist, and this twist is over-wound with copper binding wire and well soldered.

The doubled and insulated wire is then coiled up carefully into a flat circular coil of about six inches in diameter, and the stout terminal wires bent at right angles. If the wire is springy it may be tied with tape to keep it in the requisite form. We have then a coil of wire wound non-inductively and exactly six metres in length. The next step is to take the electrical resistance of this wire at zero centigrade. For this purpose the flat coil is placed in a circular glass beaker of size large enough to hold it, and the beaker is filled with paraffin oil. The beaker must be placed in a wooden tub and surrounded with broken ice. The paraffin is then kept well stirred and its temperature taken with a corrected thermometer. When the temperature is stationary the resistance of the coil is taken on the bridge. The ice is then replaced by water at different temperatures, and the resistance is again taken in like manner at the various temperatures. The paraffin oil must be kept well stirred and the temperature maintained as steadily as possible during each set of observations. These observations should be repeated several times, so as to obtain the electrical resistance of the coil at various temperatures between 0°C. and 100°C.

A correction has to be applied to these observed resistance values for the resistance of the thick copper connectors. This must be ascertained by measuring on the bridge the resistance of a loop of the same thick copper wire as that used for the connectors, and equal to them in length taken together. These resistance measurements being corrected, we have the resistance at various temperatures of a known length of the wire. The wire is then cut away from its connectors, and the silk or cotton insulation carefully removed and the wire cleaned with ether and alcohol with as little rubbing as possible. The lengths are then remeasured and are folded up as compactly as possible and weighed on the chemical balance. After weighing in air, the lengths are weighed hanging in distilled water. To get rid of the air which clings to the wire it must be gently warmed in boiled distilled water, allowed to cool under the water, and then hung by a very fine platinum wire to one scale pan, and the nett weight of the wire taken when hanging in distilled water at 15°C. If W is the weight in air and W¹ the weight in water, all corrections being applied, then the specific gravity s is given by the equation

$$s = \frac{W}{W - W^{1}}$$

If very exact results are required, a correction must be applied for the change of density of the distilled water with temperature, but the other errors of observation will generally mask this correction. One point to notice is that, in weighing the wire in water, it should be suspended by a fine platinum wire of about ·004 inch in diameter rather than by a thread. The weight of this platinum wire alone, as immersed in the water, must be taken and deducted from the gross weight. The difficulty of making a very exact weighing of a body hung in a liquid is that the capillary action of the surface layer of the liquid resists the movement of the suspending fibre through it, and thus renders the balance less sensitive.

For precautions in the use of the balance in weighing and in taking the specific gravity or density of the material, the student should consult any of the following textbooks on Practical Physics : Glazebrook and Shaw's " Practical Physics," Chapter V.; Kohlrausch's " Physical Measurements "; or Nichol's " Laboratory Manual of Physics," Vol. I., Chap. II.

The length, resistance and specific gravity being carefully determined, we can insert the observed values in the formula (3), and obtain the value of the specific volume resistance ρ of the material. From the values of the resistance at various temperatures the volume specific resistance can be calculated for these different temperatures.

The volume specific resistance in C.G.S. units should then be set out in a curve as a function of the temperature. Taking a horizontal line on which to represent degrees centigrade, we set up the values of the volume specific resistance to any scale on the vertical ordinates. For a very small range of temperature this will be found to be nearly a straight line, but for great ranges of temperature the lines (*see* Dewar and Fleming, *Philosophical Magazine*, September, 1892, p. 272) will be curved upwards or curved downwards. The line can be produced backwards to the temperature of zero centigrade and the volume specific resistance at that temperature determined. The volume specific resistance of pure copper is an important number. The various determinations by Matthiessen and others are not very closely in agreement, but the value generally called Matthiessen's Standard is as follows :—The volume specific resistance ρ of pure soft annealed copper is 1580 C.G.S. units, and the volume specific resistance of pure hard-drawn copper is 1620 C.G.S. It is very usual at the present time to meet with commercial copper wire which gives lower values than the above; that is, which has a higher conductivity than Matthiessen's Standard.

The Tables below give the results of measurements of volume specific resistance of various pure metals and alloys by Dewar and Fleming, and also the temperature coefficients (α C.) of the same. (For definition of temperature coefficient *see* Elementary Form No. 9.) The true temperature coefficient of any material at any temperature is obtained from the temperature resistance curve by taking the *slope* of the curve at that point, or the trigonometrical tangent of the angle which the geometrical tangent at that point makes with the axis of temperature. For it is obvious that, if R is the resistance at any temperature θ, then $\dfrac{dR}{d\theta}$ is the temperature coefficient at the temperature θ.

If a circular-sectioned wire of any metal has a length L centimetres and a mass of W grammes and a resistance of R ohms, then it is easily shown that if ρ^1 is the resistance in ohms of a wire of the same material, having a length of one metre and weighing one gramme, then

$$R = \frac{\rho^1 L^2}{10000 \, W},$$

or

$$\rho^1 = \frac{10000 \, W \, R}{L^2}.$$

ρ^1 is obviously the mass specific resistance as above defined. Hence, to determine this coefficient a measurement of length and resistance suffices.

Volume Specific Resistances in C.G.S. Units of Pure Metals at 0°C., and Mean Temperature Coeficients (α) between 0° and 100°C. The metals in all cases soft and annealed.			Volume Specific Resistances (ρ) in C.G.S. Units of certain Alloys at 0°C., and Temperature Coefficients (α) at 15°C.		
Metal.	ρ	α	Metal.	ρ	α
	C.G.S. Units.		Platinum-Silver	31582	0·000248
Platinum	10917	0·003669	Platinum-Iridium	30896	0·000822
Gold	2197	0·00377	Platinum-Rhodium	21142	0·00143
Palladium	10219	0·00354	Gold-Silver	6280	0·00124
Silver	1468	0·00400	Manganese-Steel	67148	0·00127
Copper	1561	0·00428	Nickel-Steel	29452	0·00201
Aluminium	2665	0·00435	German-Silver	29982	0·000279
Iron*	9065	0·00625	Platinoid	41731	0·00091
Nickel	12323	0·00622	Manganin	46678	0·0000
Tin	13048	0·00440	Silverine ....,....	2064	0·00285
Magnesium	4355	0·00981	Aluminium-Silver	4641	0·00238
Zinc	5751	0.00406	Aluminium-Copper	2904	0·00381
Cadmium	10029	0·00419	Copper-Aluminium	8847	0·000897
Lead	20390	0·00411	Copper-Nickel-Aluminium	14912	0 000645
Thallium	17693	0·00398	Titanium-Aluminium ...	8887	0·00290

* The iron here used cannot be considered as *pure*, but only as approximately pure.

Taking a sample of copper or other wire, determine, as above, its volume specific resistance and temperature coefficient at about 15°C. Enter your observations in the form on page 4.

DETERMINATION OF THE VOLUME SPECIFIC RESISTANCE AND TEMPERATURE COEFFICIENT OF A SAMPLE OF WIRE.

Nature of wire used   =
Length of wire taken   =   centimetres.
Weight of wire in air   =   grammes.
Weight of wire in water   =   grammes.
Specific gravity of wire   =
Mean diameter of wire   =   centimetres.
Mean temperature coefficient at   °C. =

Observation No.	Temperature of wire.	Resistance of wire in ohms, including connectors.	Resistance of connectors in ohms.	True resistance of wire in ohms.	Remarks.

Name *Date*

The Measurement of Low Resistances by the Potentiometer.

The apparatus required for these determinations is a potentiometer set, which may be a simple straight wire potentiometer, or a more complete form of the instrument, such as Crompton's potentiometer. The outfit includes a pair of secondary cells of at least thirty to forty ampere-hours capacity, one or more standard Clark cells, and a sensitive suspended coil galvanometer. A set of low resistances will also be required.

The Student is recommended to make a diagrammatic sketch of the arrangement of the apparatus.

A general explanation of the potentiometer and its use has been given in Elementary Form No. 10, to which the student is referred. The attention of the student must, however, be directed to the following points in connection with the use of this instrument.

The Insulation of the Instrument.—Difficulties are sometimes experienced in the use of the potentiometer in very accurate work, and when an exceedingly sensitive galvanometer is being used, owing to leakage currents passing through the galvanometer. In some cases merely touching with the hand certain parts of the circuit is sufficient to make galvanometer deflections which altogether mask the real potential differences to be observed. The first point to notice is that the potentiometer, the working secondary cell or cells and the galvanometer should all be insulated by placing them on slips of clean ebonite. The connecting wires should be insulated with gutta percha or indiarubber, and in some cases the observer should stand on a sheet of indiarubber or ebonite and have the finger with which he presses the slide key down insulated with a rubber finger-stall.

Steadiness of Working E.M.F.—In the next place it is absolutely essential for good work to have a very steady source of E.M.F. to supply the working current. This can only be obtained by using rather large secondary cells which have been well charged and then *partly discharged* before using for the potentiometer. The secondary cells give the steadiest E.M.F. when about 25 per cent. of their full quantity has been taken out. If a freshly-charged cell is used it is liable to sudden small changes of electromotive force.

Galvanometer.—The sensitiveness of the potentiometer ultimately depends upon the galvanometer employed with it. If no dynamos or other magnetic bodies are near, a suspended needle galvanometer, such as a high resistance mirror galvanometer, can be used with great advantage ; but if used in a factory this is out of the question, from want of steadiness and on account of disturbing magnetic fields. The dead-beatness and independence of external fields greatly recommend the suspended coil galvanometer. The general conditions with which a sensitive galvanometer for this purpose should comply are, however, that one hundred-millionth of an ampere passing through it should cause a deflection of the spot of light of at least one millimetre when the scale is at a distance of one metre from the mirror. This may be regarded as a good, but by no means an unusual, degree of sensitiveness. The sensitiveness of a galvanometer can be conveniently defined by stating the current through it which will create a deflection of one minute of an angle of its coil or needle. The galvanometer circuit should always have, either in the galvanometer or outside, a resistance of at least 1,000 ohms, in order that the Clark cell may never send a current of more than a very small amount, even in the process of finding the balance.

CLARK CELLS.—It has been abundantly demonstrated that when Clark cells are made up with pure materials in a certain manner, they preserve for long periods of time a perfect constancy of electromotive force.

The following is the Board of Trade specification for setting up Clark cells :—

ON THE PREPARATION OF THE CLARK CELL.

Definition of the Cell.

The cell consists of zinc or an amalgam of zinc with mercury and of mercury in a neutral saturated solution of zinc sulphate and mercurous sulphate in water, prepared with mercurous sulphate in excess.

Preparation of the Materials.

1. *The Mercury.*—To secure purity it should be first treated with acid in the usual manner, and subsequently distilled in vacuo.

2. *The Zinc.*—Take a portion of a rod of pure redistilled zinc, solder to one end a piece of copper wire, clean the whole with glass paper or a steel burnisher, carefully removing any loose pieces of the zinc. Just before making up the cell, dip the zinc into dilute sulphuric acid, wash with distilled water, and dry with a clean cloth or filter paper.

3. *The Mercurous Sulphate.*—Take mercurous sulphate, purchased as pure, mix with it a small quantity of pure mercury, and wash the whole thoroughly with cold distilled water by agitation in a bottle; drain off the water and repeat the process at least twice. After the last washing, drain off as much of the water as possible.

4. *The Zinc Sulphate Solution.*—Prepare a neutral saturated solution of pure (" pure recrystallised ") zinc sulphate by mixing in a flask distilled water with nearly twice its weight of crystals of pure zinc sulphate, and adding zinc oxide in the proportion of about 2 per cent. by weight of the zinc sulphate crystals to neutralise any free acid. The crystals should be dissolved with the aid of gentle heat, but the temperature to which the solution is raised should not exceed 30°C. Mercurous sulphate treated as described in 3 should be added in the proportion of about 12 per cent. by weight of the zinc sulphate crystals to neutralise any free zinc oxide remaining, and the solution filtered, while still warm, into a stock bottle. Crystals should form as it cools.

5. *The Mercurous Sulphate and Zinc Sulphate Paste.*—Mix the washed mercurous sulphate with the zinc sulphate solution, adding sufficient crystals of zinc sulphate from the stock bottle to insure saturation, and a small quantity of pure mercury. Shake these up well together to form a paste of the consistence of cream. Heat the paste, but not above a temperature of 30°C. Keep the paste for an hour at this temperature, agitating it from time to time, then allow it to cool ; continue to shake it occasionally while it is cooling. Crystals of zinc sulphate should then be distinctly visible, and should be distributed throughout the mass. If this is not the case, add more crystals from the stock bottle, and repeat the whole process.

This method insures the formation of a saturated solution of zinc and mercurous sulphates in water.

To set up the Cell.

The cell may conveniently be set up in a small test tube of about 2cm. diameter and 4cm. or 5cm. deep. Place the mercury in the bottom of this tube, filling it to a depth of, say, 0·5cm. Cut a cork about 0·5cm. thick to fit the tube ; at one side of the cork bore a hole through which the zinc rod can pass tightly ; at the other side bore another hole for the glass tube which covers the platinum wire : at the edge of the cork cut a nick through which the air can pass when the cork is pushed into the tube. Wash the cork thoroughly with warm water, and leave it to soak in water for some hours before use. Pass the zinc rod about 1cm. through the cork.

Contact is made with the mercury by means of a platinum wire about No. 22 gauge. This is protected from contact with the other materials of the cell by being sealed into a glass tube. The ends of the wire project from the ends of the tube ; one end forms the terminal, the other end and a portion of the glass tube dip into the mercury.

Clean the glass tube and platinum wire carefully, then heat the exposed end of the platinum red hot, and insert it in the mercury in the test tube, taking care that the whole of the exposed platinum is covered.

Shake up the paste and introduce it without contact with the upper part of the walls of the test tube, filling the tube above the mercury to a depth of rather more than 1cm.

Then insert the cork and zinc rod, passing the glass tube through the hole prepared for it. Push the cork gently down until its lower surface is nearly in contact with the liquid. The air will thus be nearly all expelled, and the cell should be left in this condition for at least 24 hours before sealing, which should be done as follows :—

Melt some marine glue until it is fluid enough to pour by its own weight, and pour it into the test tube above the cork, using sufficient to cover completely the zinc and soldering. The glass tube containing the platinum wire should project some way above the top of the marine glue.

The cell may be sealed in a more permanent manner by coating the marine glue, when it is set, with a solution of sodium silicate, and leaving it to harden.

The cell thus set up may be mounted in any desirable manner. It is convenient to arrange the mounting so that the cell may be immersed in a water bath up to the level of, say, the upper surface of the cork. Its temperature can then be determined more accurately than is possible when the cell is in air.

In using the cell sudden variations of temperature should, as far as possible, be avoided. The form of the vessel containing the cell may be varied. In the II form, the zinc is replaced by an amalgam of 10 parts by weight of zinc to 90 of mercury. The other materials should be prepared as already described. Contact is made with the amalgam in one leg of the cell, and with the mercury in the other, by means.of platinum wires sealed through the glass.

Several cells should be set up according to this specification and tested against each other on the potentiometer. For very accurate work the observer should determine for himself the temperature coefficient of the cell he is using, which can always be done when there are two cells.

In ordinary use he can assume the values for the E.M.F. at different temperatures given in Elementary Form No. 10.

CONSTRUCTION OF LOW RESISTANCE STANDARDS.—In order to employ the potentiometer for the measurement of low resistances, it is necessary to be provided with certain standards of low resistance, particularly with a resistance of 0·1, 0·01, 0·001 ohm, the first of which will carry without sensible heating ten amperes, the second a hundred amperes, and the third a thousand amperes. It is a comparatively easy matter to copy with the potentiometer a low resistance having a given standard. To measure a resistance with the potentiometer, the resistance is joined up in series with a known low resistance standard, say of 0·1 or 0·01 ohm. Through these is sent the current from a secondary cell or cells, which must be sufficient to make a fall of pressure down the low resistances of not less than 0·1 of a volt. From the ends of the known low resistance and from the ends of the unknown low resistance potential wires are taken to the potentiometer, and the potential difference between these points is quickly compared. Since the same current flows through both resistances, the magnitude of the resistances are proportional to the fall of potential down them. Hence, if one resistance has a known value that of the other resistance becomes known. If, instead of having to measure an unknown low resistance, we have to construct one, the following method may be adopted:—The first problem is to construct a resistance, say, of 0·1 of an ohm, and which will carry 10 amperes without sensible heating. This may be achieved by taking bare platinoid or manganin wire of No. 16 S.W.G. size and cutting off 10 lengths, which are each rather more than one ohm in resistance.

To find out what length of a given wire will have one ohm resistance, measure the resistance of one yard or metre of it on the Wheatstone's Bridge, and then calculate the length required to give one ohm. Cut off ten lengths rather greater than one ohm in resistance. These ten one-ohm wires are all then soldered at one end to a broad thick plate of high conductivity copper, and a final and accurate adjustment made by gradually snipping over bits of each wire, until each wire measures exactly one ohm up to a mark one centimetre from the end. When this is the case, the free ends are all soldered on to a similar block of copper, the overlap of wire being one centimetre. The loose wires can then be tied together, and terminal screws placed on the copper blocks. We have then a resistance of 0·1 ohm made by joining ten ohm wires in parallel. By the method above described this tenth ohm can be copied. A strip of manganin is taken of width and thickness sufficient to carry ten amperes without sensible heating, and of a resistance rather greater than 0·1 ohm. Terminal screws are provided at each end, and a little way from one end a thin potential wire is soldered to the strip. This strip is then placed in series with the known tenth-ohm, and by moving a loose potential wire along it we can find the point where the fall of volts down it is one volt for ten amperes. At that point the second potential wire is soldered and the resistance strip mounted on a convenient frame, with current and potential terminals.

Low resistances can be compared together with moderate accuracy by the method of direct deflection of the galvanometer. The low resistances (known and unknown) are connected in series, and the current from one or two large dry cells or secondary cells sent through them. The wires from a movable coil galvanometer are then connected to the ends of the resistances successively, and the direct deflections of the galvanometer noted. These deflections are proportional to the resistances if we assume that the deflections are proportional to the currents through the galvanometer. The method is not applicable for very refined work, but is useful in measuring the resistance of secondary circuits of transformers. Taking some known and unknown low resistances, the student should practise the comparison of them as above described, and record his observations on the Table over leaf.

THE MEASUREMENT OF LOW RESISTANCES BY· THE POTENTIOMETER.

Observation No.	Temperature of Clark cell.	Scale reading on wire for Clark cell.	Scale reading on wire for known resistance.	Value of known resistance.	Scale reading on wire for unknown resistance.	Calculated value of unknown resistance.	Remarks.

ELECTRICAL LABORATORY NOTES AND FORMS.

No. 23.—ADVANCED (No. 3).

Name   Date

The Measurement of Armature Resistances.

The apparatus required for these experiments is a high resistance movable coil galvanometer, a potentiometer set, two or three secondary cells, and some standard low resistances, 0·1 ohm and 0·01 ohm. A series of dynamos must be at hand on which to experiment.

The Student is recommended to carefully sketch the arrangement of the circuits and apparatus.

The armature of a dynamo or alternator consists of a loop of wire or copper bars so disposed that it forms practically either a single loop conductor or else a ring conductor with connections at diametrically opposite ends. The armatures of dynamos are made of low resistance in order that as small a proportion as possible of the mechanical energy transformed into electric current energy may be wasted in them. This resistance is generally something of the order of a tenth or one-hundredth of an ohm. Such a low resistance cannot be accurately measured on the Wheatstone's Bridge, because the variable error introduced by the resistances of the contacts would frequently be greater than the resistance to be measured. It can, however, easily be measured by the method of deflections or by the potentiometer.

The arrangements required for this purpose are as follows :—Obtain, in the first place, an approximate measure of the resistance of the armature by the method of galvanometer deflections. Take a known low resistance of one-hundredth of an ohm in the form of a strip resistance. Carefully clean the commutator or collector rings of the dynamo, and place the brushes on it. If the dynamo is a continuous-current dynamo, take some pains to see that the brushes are pressing against exactly opposite segments on the commutator. If the field magnet leads are taken off directly from the brushes, these leads must be removed or disconnected for the time being. Join up in series the standard low resistance, the armature to be measured, one or two rather large secondary cells and a plug key, so that the current from the cells can flow through the armature and through the standard low resistance successively. From the ends of the known low resistance take off a pair of potential wires. Join also potential wires to the ends of the armature. This is best done by pressing the flattened and cleaned ends of the potential wires underneath the brushes and thus causing them to press against the commutator segments or collector rings. These two pairs of potential wires are then brought to four mercury cups placed near the galvanometer.

The galvanometer should be set up in a steady place and have a uniformly divided scale. The image reflected on to the scale should be the image of a fine wire illuminated from behind, so as to get scale readings of the deflection of the coil as sharply as possible. When all is ready the plug key is closed, so as to send the current through the armature and standard resistance. The potential wires from the ends of the armature and resistance are to be connected alternately to the galvanometer. It will usually be necessary to add a high resistance in series with the galvanometer to reduce the deflection to a convenient amount, but it should be as large as possible, and it should be possible to read this deflection to at least one per cent. of its value. Let the deflection be noted and called d_2, when the potential wires from the ends of the known low resistance are connected to the galvanometer. In the same way let the deflection be noted and called d_1 when the potential wires from the ends of the armature are connected to the galvanometer. It is well to take the deflection due to the fall of potential down the known low resistance *before* and *after* that due to the armature, and then to take the mean of these two results. In this way we eliminate any error due to the sinking of the electromotive force of the battery during the experiment. These deflections d_1 and d_2 are proportional, or nearly so, to the currents flowing through the galvanometer in the two cases. When the galvanometer is connected to the ends of the standard resistance, or to those of the armature, a current flows through the galvanometer and associated high resistance, which is numerically equal to the quotient of this difference of potential.by the total resistance of the galvanometer current. Let R_1 be the unknown resistance of the armature in ohms, and let V_1 be the fall in potential in volts down it when the steady current is flowing through it. In the same way, let R_2 be the resistance of the known low resistance standard, and let V_2 be the fall in potential down it. Then the current flowing through the armature is equal to $\frac{V_1}{R_1}$ amperes, and the current flowing through the standard resistance is $\frac{V_2}{R_2}$. But these currents must have the same value because the standard resistance and armature are in series. Hence,

$$\frac{V_1}{R_1} = \frac{V_2}{R_2}, \text{ or } \frac{V_1}{V_2} = \frac{R_1}{R_2}.$$

Again, if r be the resistance of the galvanometer and its associated high resistance : then, when the potential wires from the ends of the known low resistance are put in connection with the galvanometer, the current through the galvanometer is equal to $\frac{V_2}{r}$ amperes, and if the deflection of the galvanometer is d_2, and if C is the galvanometer constant, we have

$$\frac{V_2}{r} = C\, d_2.$$

In the same way, if d_1 is the deflection when the potential wires from the ends of the armature are put in connection with the galvanometer, we have

$$\frac{V_1}{r} = C\, d_1.$$

Hence

$$\frac{V_1}{V_2} = \frac{d_1}{d_2};$$

but we have found that

$$\frac{V_1}{V_2} = \frac{R_1}{R_2}.$$

Hence

$$\frac{R_1}{R_2} = \frac{d_1}{d_2},$$

or

$$R_1 = R_2 \frac{d_1}{d_2}.$$

We know the value of R_2, and we know the value of d_1 to d_2; hence we can calculate the value of R_1. In the above method it is assumed that the deflections of the galvanometer are exactly proportional to the currents producing them. This is not always or even generally exactly true. Hence the method can be improved by using the potentiometer to measure the fall of potential V_1 and V_2 down the armature and standard resistance. If this is done, considerably greater accuracy in measurement may result, but the method of galvanometer deflection will in general be accurate enough for most laboratory work; and we can always, if need be, calibrate the galvanometer scale so as to ascertain what the deflections really mean. In one of these two ways the student should practise measuring the resistance of several armatures, and also of the low tension coil of several transformers, recording this result in the appended form.

THE MEASUREMENT OF ARMATURE RESISTANCES.

$R_2 = $ Standard low resistance used =   ohm.

Observation No.	Galvanometer deflection or potential difference at ends of standard resistance. d_2 or V_2.	Galvanometer deflection or potential difference at ends of armature. d_1 or V_1.	Calculated value of the unknown resistance R_1. $R_1 = R_2 \dfrac{d_1}{d_2}$, or $R_1 = R_2 \dfrac{V_1}{V_2}$.	Description of armature or coil measured.

ELECTRICAL LABORATORY NOTES AND FORMS.

No. 24.—ADVANCED (No. 4).

Name   *Date*

Standardization of an Ammeter by Copper Deposit.

The apparatus required for these experiments is a good chemical balance, which must be capable of weighing a mass of 100 grammes with an accuracy of 0·1 of a milligramme, a large copper depositing cell as described below, a source of constant current, some regulating resistances, and instruments to calibrate.

The Student is recommended to sketch the arrangement of the circuits and apparatus.

The passage of a current of electricity through a solution of copper sulphate by means of copper electrodes causes copper to be dissolved off one plate or electrode and deposited on the other. The first plate is spoken of as the *loss plate* and the second as the *gain plate*. If certain precautions be taken, as stated below, the amount of copper deposited on the gain plate in a second bears a definite and fixed relation to the average strength of the current passing through the electrolyte. In exact researches there are certain advantages to be gained by the use of a solution of silver nitrate, and silver plates instead of copper sulphate and copper plates, and the British Board of Trade have adopted the silver deposition as a basis for the practical determination of current strength.

The legal definition of the ampere is as follows :—One ampere is the denomination of a current of unvarying strength which deposits silver at the rate of 0·001118 of a gramme per second.

The weight of metal deposited by one ampere per second is called the electro-chemical equivalent of the metal.

The following are the directions given by the Board of Trade for determining the strength of a current by the method of silver deposit :—

In the following specification the term silver voltameter means the arrangement of apparatus by means of which an electric current is passed through a solution of nitrate of silver in water. The silver voltameter measures the total electrical quantity which has passed during the time of the experiment; and by noting this time, the time average of the current, or, if the current has been kept constant, the current itself can be deduced.

In employing the silver voltameter to measure currents of about one ampere, the following arrangements should be adopted. The cathode on which the silver is to be deposited should take the form of a platinum bowl not less than 10cm. in diameter, and from 4cm. to 5cm. in depth.

The anode should be a plate of pure silver some 30 sq. cm. in area and 2mm. or 3mm. in thickness.

This is supported horizontally in the liquid near the top of the solution by a platinum wire passed through holes in the plate at opposite corners. To prevent the disintegrated silver which is formed on the anode from falling on to the cathode, the anode should be wrapped round with pure filter paper, secured at the back with sealing wax.

The liquid should consist of a neutral solution of pure silver nitrate, containing about 15 parts by weight of the nitrate to 85 parts of water.

The resistance of the voltameter changes somewhat as the current passes. To prevent these changes having too great an effect on the current, some resistance besides that of the voltameter should be inserted in the circuit. The total metallic resistance of the circuit should not be less than 10 ohms.

Method of making a Measurement.

The platinum bowl is washed with nitric acid and distilled water, dried by heat and then left to cool in a desiccator. When thoroughly dry it is weighed carefully.

It is nearly filled with the solution, and connected to the rest of the circuit by being placed on a clean copper support to which a binding screw is attached. This copper support must be insulated.

The anode is then immersed in the solution so as to be well covered by it and supported in that position ; the connections to the rest of the circuit are made.

Contact is made at the key, noting the time of contact. The current is allowed to pass for not less than half-an-hour, and the time at which contact is broken is observed. Care must be taken that the clock used is keeping correct time during this interval.

The solution is now removed from the bowl, and the deposit is washed with distilled water and left to soak for at least six hours. It is then rinsed successively with distilled water and absolute alcohol, and dried in a hot-air bath at a temperature of about 160 C. After cooling in a desiccator it is weighed again. The gain in weight gives the silver deposited.

To find the current in amperes, this weight, expressed in grammes, must be divided by the number of seconds during which the current has been passed, and by 0·0011118.

The result will be the time-average of the current, if during the interval the current has varied.

In determining by this method the constant of an instrument the current should be kept as nearly constant as possible, and the readings of the instrument observed at frequent intervals of time. These observations give a curve from which the reading corresponding to the mean current (time-average of the current) can be found. The current, as calculated by the voltameter, corresponds to this reading.

For most purposes the deposit of copper can be made to give results of almost equal exactness with silver, provided that certain precautions are taken.

To determine the strength of a steady or unvarying current by the method of copper deposit, and to check an ammeter reading at the same time, the following arrangements are made :—

A glass cylindrical vessel is taken and filled nearly to the top with a slightly acid solution of sulphate of copper. In this solution are suspended an odd number of copper plates, alternate plates being connected together. This vessel is called a copper voltameter. The size of plates, and therefore of the voltameter cell, is determined by the strength of the current it is desired to measure, by the rules given below. The 1st, 3rd, 5th, 7th, &c., plates are metallically connected, and these plates are made the loss plates. The 2nd, 4th, 6th, &c., plates are also metallically connected, and are made the gain plates. If there are N gain plates and each has a total surface area of S square centimetres, reckoning both sides, then N S square centimetres is the total area of gain plates ; and, similarly, if S is the total surface of the single loss plate, and there are N + 1 loss plates, the total opposed surface of the loss plates is also N S square centimetres. The total surface of loss plates should never be less than 40 square centimetres per ampere. If the area is much less than this amount the resistance of the cell becomes high and variable, owing to the formation of copper oxide on the plate. The gain plates should never expose a less area than 20 square centimetres per ampere, or else the copper deposit is not firmly adherent to the plate, and a portion of it may be lost in washing or drying. It is always safest to employ from 50 to 100 square centimetres of plate surface per ampere. The size of electrolytic cell or voltameter is therefore determined by the currents to be measured. Assuming that a current, say, of about five amperes is to be measured, proceed as follows :—The current must be supplied from secondary cells, and be regulated by a rheostat or resistance capable of being varied by exceedingly small steps. One of the most convenient forms is a carbon plate rheostat, in which the resistance is varied within certain limits by squeezing more or less tightly a series of carbon plates, about 3 inches square, held in a frame. The current passes through the ammeter or ampere-balance to be checked, and then through a voltameter with copper plates of such size and number that we have about 50 square centimetres of surface per ampere.

The copper plates must be so held in this cell that they cannot move during the experiment, but be capable of being easily taken out of the cell for the purpose of weighing. Details of the best way of doing this will be found in a Paper by Mr. A. W. Meikle, read before the Physical Society of Glasgow University, January 27th, 1888, or in one by Mr. Thomas Gray, published in the *Philosophical Magazine* for November, 1886.

The copper sulphate solution must be made by dissolving pure crystals of the salt in distilled water until a solution of a density of 1·15 to 1·18 is obtained. One per cent. by volume of free strong sulphuric acid must then be added, and this addition of free acid is essential to success. No good results can be obtained with neutral solutions. The copper plates must be cleaned by dipping them in strong nitric acid and washing and drying carefully. They ought to be perfectly bright and clean, and not afterwards touched with the fingers. Each plate should have a number stamped on it for recognition. The edges of the plates and sharp angles should be first rounded off with the file. Each plate must then be carefully weighed. Before weighing, the plates must be kept for some time in a desiccating chamber over strong sulphuric acid to remove the film of moisture adhering to them. The plates, when weighed, are then to be arranged in the voltameter and connected up in the proper manner.

The voltameter is connected in series with the ammeter to be checked, and at a known instant of time the current is started through the cell. The observer has then to keep the current as constant as possible by means of the variable carbon resistance, and this constant current must be kept flowing through the cell for some hours. The duration of the experiment must be observed by a good clock or watch, as the absolute determination of the time is involved. At the same time the scale reading of the ammeter to be checked is carefully taken. When a sufficient time has elapsed the current is stopped and the plates removed from the electrolytic cell The gain plates should be taken out at once and be given a rinse in water to free them from adhering solution. A few drops of sulphuric acid must be added to this water to prevent oxidation. The plates are then dried on clean white blotting paper, and put into a copper hot air oven for a few minutes to dry completely. They are then transferred to the desiccator to cool and finally weighed. The exact gain of each plate is noted, and the total gain obtained by addition. The gain in weight per second is then calculated and divided by the electrochemical equivalent of copper, and the resulting quotient is the value of the mean or average current in amperes. The electrochemical equivalent of copper varies very slightly with current density and temperature as follows :—

Area of gain plates in square centimetres per ampere.	Electrochemical equivalent of copper	
	at 12°C.	at 23°C.
50	0·0003287	0·0003286
100	0·0003284	0·0003283
150	0·0003281	0·0003280
200	0·0003279	0·0003277
250	0·0003278	0·0003275
300	0·0003278	0·0003272

The Student should make in this way a careful check of the reading of an ammeter or current-balance at one or two points on the scale. The results should be entered up in the form on the next page.

(4)

STANDARDIZATION OF AN AMMETER BY COPPER DEPOSIT.

Observation No.	Number of plate.	Weight of (gain) copper plates at beginning.	Weight of (gain) copper plates at end.	Total gain in weight of plates.	Duration of experiment.	Calculated mean value of the current.	Reading of ammeter or galvanometer.

ELECTRICAL LABORATORY NOTES AND FORMS.

No. 25.—ADVANCED (No. 5).

· *Name* ...   *Date·*

The Standardization of a Voltmeter by the Potentiometer.

The apparatus needed for these tests is a complete potentiometer set, preferably Crompton's form of the instrument, and a divided resistance consisting of a coil or coils of wire divided in known and fixed ratios. The galvanometer employed with the potentiometer should be a very sensitive movable coil galvanometer, and the Clark cells used should have been checked against reliable standard cells for comparison.

The Student should sketch the arrangement of apparatus as set up.

The general arrangement of the simple wire potentiometer has been fully described in Elementary Form No. 10. For most exact work it is better to employ the more complete form of instrument arranged by Crompton. In this instrument only a part of the resistance wire is actually a stretched wire; the major part consists of coils of wire wound on bobbins placed underneath the base of the instrument.

A more detailed description of this potentiometer may be given with the help of Fig. 1 (*see* page 2).

A B is a short stretched manganin wire, and one end of it is connected with 14 exactly similar coils of wire, joined in series, each of which measures about two ohms. These 14 coils, marked 1 to 14 E, are arranged so that the contiguous ends of each coil are brought to a set of 14 studs arranged in a semicircle on the board. The ends of this resistance, consisting of the 14 coils of wire and the straight stretched wire equal in resistance to one of them, constitute the potentiometer wire, and it has a total resistance of about 30 ohms. The end of this series is connected to another set of 14 coils of wire, G, joined in series, each equal to two ohms in resistance and arranged with contiguous ends joined to 14 studs on the board, and the last coil is connected to a spiral sliding resistance or rheostat G_1, so that a graduation of resistance may be made. To the ends of this complete resistance is joined one cell of a secondary battery, which constitutes the working cell of the potentiometer. The connections are made as in the Diagram in Fig. 1 in (No. 6) Advanced Form.

By turning the switch handle G to rest on different studs of the set of 14 coils, and varying the sliding resistance G_1 by turning the disc, we can introduce resistances varying from zero to about 30 ohms in series with the other 14 coils E and stretched wire A B. By this means we can always make the potential difference between the ends of the resistance consisting of the 14 coils and straight wire exactly equal to 1·5 volts. The straight manganin wire lies on a scale divided into 1,000 parts. Hence, when the fall of potential down the wire and coils is 1·5 volts, the fall down a length of the wire equal to one of the scale divisions is one ten-thousandth of a volt. Along and over the wire moves a slider, C, which can make a contact with the wire at

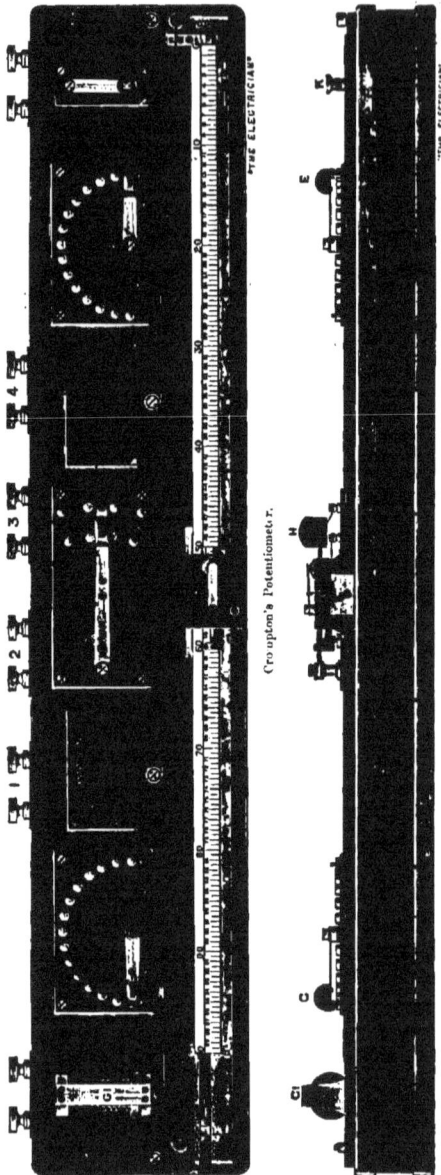

Fig. 1.

Cropton's Potentiometer.

any point. One terminal of a galvanometer is attached to the centre of the radial contact arm E, which works over the studs of the 14 coils so as to make a contact at any stud, and the other end is joined to one terminal of a double-pole radial switch H, the second arm of which is joined to the sliding contact piece on the manganin wire. This radial switch can be moved over to make double contacts with a set of four or six pairs of studs, and these studs are in connection with a set of four or six pairs of terminals on the front of the base board marked 1 to 4, by means of which any cell or source of electromotive force may be inserted in series with the galvanometer and slider.

We have then also to provide a pair of standard Clark cells, made as described in (No. 2) Advanced Form, and a very sensitive movable coil galvanometer. For voltmeter checking we have in addition to provide a divided resistance or volt box. This consists of a coil of manganin wire of about 5,000 or 10,000 ohms in resistance, and which is divided in certain exact ratios, so that one section is, say, 50 ohms, and another 100, and another 500; so that we have a resistance which will bear a terminal potential difference of 100 to 150 volts put upon it. We are then able to make contacts through the terminals with points on this resistance between which there is $\frac{1}{100}$th, $\frac{1}{50}$th, or $\frac{1}{10}$th of the potential difference there is between the ends of the whole resistance.

To check and calibrate a voltmeter with this apparatus we proceed as follows: Suppose the voltmeter reads from 60 to 100 or 120 volts. Provide a set of 50 small secondary cells. For this purpose none are more convenient than the small lithanode glass tube secondary cells, but they must be arranged so that any number of cells can be employed.

Join up the volt box or divided wire across the terminals of the voltmeter to be tested, and connect, say, 50 cells across the terminals. The voltmeter will now read nearly 100 on its scale. To find out what the potential difference really is, bring wires from the $\frac{1}{100}$th section of the volt box resistance to the potentiometer, and connect them in, as described in Elementary Form No. 10, to one of the pairs of terminals; join in also a couple of Clark cells to any other two pairs of terminals on the potentiometer. See that the potentiometer and galvanometer are all carefully insulated, and begin by setting the potentiometer by the Clark cell by varying the resistance of the series of twelve coils and spiral resistance, so that the fall of potential down the 14 coils and slide wire is exactly 1·434 volts for 14340 scale reading ; that is, if the E.M.F. of the Clark cell is 1·434 volts at that time, then the Clark cell must be made to balance at 14340 on the scale. This is achieved by setting the radial arm E of the 14 coils to touch stud 14 and setting the slider on the manganin wire to touch at 340 on the scale of the slide wire, and then varying the resistance of the other coils G and spiral resistance G_1 until the galvanometer shows no deflection. When this is the case the fall of potential down the slide wire is ·0001 volt per division.

The two Clark cells should then be compared, and, if in agreement, the double-pole working arm H should be switched over so as to throw in one hundreth part of the potential difference on the terminals of the voltmeter. The switch arm of the 14 coils will now have to be moved over to another stud on series E, say to stud 9 or stud 10, and a position of galvanometer balance found by moving the slider. Suppose the balance is found with arm on stud 10 and slider at 125. This indicates that the potential difference at terminals of voltmeter is 101·25. If the voltmeter reads 100 on its scale, the voltmeter error is + 1·25 volts at 100. In this way, by altering the number of battery cells, the voltmeter may be checked all along its scale. The student should make an accurate check in this way of one or more voltmeters, and set out a curve of errors by drawing vertical lines above or below a horizontal datum line (which lines are equal on some scale to the + or − errors), and setting these lines at points on the equally divided horizontal line chosen to represent the scale divisions of the voltmeter. The results should be entered up in the form on page 4.

STANDARDIZATION OF A VOLTMETER.

Observation No.	Temperature.	Value of Clark cell for that temperature.	Scale reading on potentiometer for fraction of voltmeter potential.	Divided resistance used.	True value of potential difference of voltmeter terminals.	Observed voltmeter reading.	Error of voltmeter.

ELECTRICAL LABORATORY NOTES AND FORMS.

No. 26.—ADVANCED (No. 6).

Name   *Date*

The Standardization of an Ammeter by the Potentiometer.

The apparatus required for these experiments is a complete potentiometer set (Crompton's form) as described in (No. 5) Advanced Form and, in addition, a set of low resistance standards prepared or copied from known standards as in (No. 2) Advanced Form. Arrangements must be made for taking the required current from large secondary cells and regulating it by resistance as required.

The Student is recommended to sketch the arrangement of apparatus as set up.

To standardize an ammeter by the potentiometer, this last instrument is set up and adjusted in accordance with the instructions given in Advanced Forms (No. 2) and (No. 5), which the student is assumed to have read. A diagram of the connections is shown in Fig. 1.

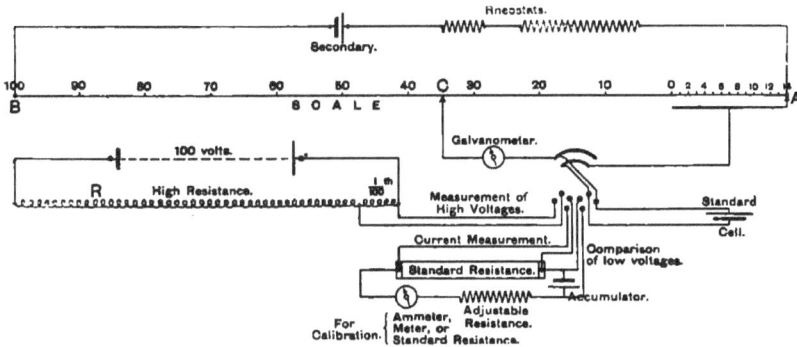

Fig. 1.

The ammeter to be calibrated is set up and joined in series with a known low resistance standard, which will carry currents of the strength required without sensible heating. Thus, if the ammeter reads from 0 to 10 amperes, we should select a 0·1 ohm standard, which will carry 10 amperes without sensible heating. These low resistance standards are best made of gilt manganin strip. The ammeter and resistance are joined in series with a carbon rheostat, which consists of plates of carbon about 3 or 4 inches square, and which can be more or less compressed with a screw. These three pieces of apparatus—ammeter, carbon resistance and standard resistance—should be connected to one or more large cells of a secondary battery sufficient to provide a steady current of the maximum magnitude necessary. This being done, the ends of the low resistance standard are joined by potential wires to the potentiometer, care being taken to send the current through the standard in the right direction for fall of potential (*see* Elementary Form No. 11.). A Clark cell is connected to the potentiometer and the potentiometer set by it. We then measure the fall of potential down the low resistance standard, and read at the same time the ammeter indication. The current through the ammeter is then changed to another value and the same measurement repeated. Suppose the fall of potential down the 0·1 ohm standard is found to be ·951 volt, and the reading of the ammeter at that instant is 10·2, this indicates that the true current through it is 9·51, and the scale error is therefore + ·69. The student should in this way check a few ammeters and record the scale errors at various points of the scale.

In thus checking an ammeter it is always necessary to reverse the current and check it at all points of the scale, both with increasing currents and with decreasing currents, because some types of ammeter in which soft iron is employed are quite different in their reading under these conditions. A good ammeter should comply with the following requirements :—

(*i.*) It should be as dead beat as possible ; that is, the needle should come very quickly to rest when the current is sent through it.

(*ii.*) It should give the same scale reading for the same current whether that scale reading is reached by increasing from a smaller current or decreasing from a larger one.

(*iii.*) It should not be affected by external magnetic fields.

(*iv.*) It should be sensitive ; that is, a very small change in the current should show itself immediately on the scale reading of the instrument.

(*v.*) Other desirable requirements are that the scale divisions should be equal in magnitude throughout the scale, and that it should begin to read from the zero point, and not have a blank space of non-useful scale.

The above requirements should also be fulfilled by a good voltmeter, and in addition this last should have a negligible or known coefficient of temperature correction. These conditions are not always fulfilled by instruments in which soft iron cores are employed which are moved in magnetic fields. Hence the above tests should always be applied in examining any instrument before passing an opinion upon its merits as an ammeter.

The student will find it a useful exercise to take some good form of ammeter or current balance and check it at some point on the scale, both by the copper deposit method, as described in Advanced Form (No. 4), and also to check it by the potentiometer method as above described, and see how far these entirely different physical methods lead to the same result.

By the use of the potentiometer all measurements made in the electrical laboratory with continuous currents may be ultimately reduced or referred to a standard of electromotive force represented by a Clark cell and a standard resistance.

The student should enter up his observations in checking any ammeters in the appended form.

STANDARDIZATION OF AN AMMETER.

Observation No.	Temperature.	Value of E.M.F. of Clark cell at that temperature.	Potentiometer reading or potential difference at ends of standard resistance.	Value of standard resistance employed.	True value of current in amperes.	Scale reading of ammeter.	Error of ammeter.

STANDARDIZATION OF AN AMMETER.

Observation No.	Temperature.	Value of E.M.F. of Clark cell at that temperature.	Potentiometer reading or potential difference at ends of standard resistance.	Value of standard resistance employed.	True value of current in amperes.	Scale reading of ammeter.	Error of ammeter.

ELECTRICAL LABORATORY NOTES AND FORMS.

No. 27.—ADVANCED (No. 7).

Name   *Date*

Determination of the Magnetic Permeability of a Sample of Iron.

The apparatus required for these experiments is a set of soft iron rings, circular in section, and of which the diameter of cross-section is not large compared with the mean diameter of the ring. A ballistic galvanometer and condenser for standardizing it is required, and also a large soft iron ring cut into two parts.

The Student is recommended to sketch the arrangement of the apparatus as set up for this test.

If a very long rod of iron is placed in a long magnetising coil, or solenoid, and is magnetised by a uni-directional current sent through the coil, the magnetic force in the interior of that solenoid can be calculated, as shown in Elementary Form No. 7. If the iron rod is, however, a short one—that is to say, if it is less than about 50 diameters long—the magnetic poles induced in the ends of the bar make their effect felt in diminishing the magnetic force due to the coil in the space occupied by the rod. Assuming the rod, however, to be a long one—that is to say, 100 or more diameters long—we can calculate the magnetic force H, due to a current of A amperes circulating in the solenoid of N turns and length L, by the formula—

$$H = \frac{4\pi}{10} \frac{A\,N}{L}.$$

If a secondary circuit of wire is wound closely round the centre of the iron rod, and if the current in the magnetising solenoid is reversed suddenly—the secondary circuit being connected with a ballistic galvanometer—it is shown in Elementary Forms Nos. 7 and 8 that we can calculate the induction density B in the iron at the place where the secondary coil is wound. The magnetic induction density B, or, as it is commonly called, the number of lines of force per square centimetre, and the magnetic force H, are related to one another by the equation—

$$B = \mu\,H,$$

where μ is the *magnetic permeability* of the material. The object of the following experiments is to determine the magnetic permeability of a sample of iron. Whenever that iron can be furnished in the form of a ring, the determination is very simple. The iron ring should be turned up in the lathe and formed so that it has a circular cross-section of radius a and a mean diameter of $2r$. These measurements should be taken in six different places and the mean value obtained. The ring may be, preferably, about one centimetre in diameter of cross-section and ten centimetres in mean diameter. The ring should then be carefully wound over with N turns of insulated wire in one layer

No. 18 S.W.G., double cotton or silk-covered copper wire being used, and the mean diameter of the cross-section again taken over the wire. Let this be $2b$ centimetres. Then $\frac{a+b}{2}$ is the mean radius of each circular turn of wire put on the ring. Let there be N turns of wire in all on the ring. Then the mean magnetic force H inside the wire windings can be calculated, and is equal to

$$\frac{4}{10}\ \frac{N A}{a^2}\ \left\{ r - \sqrt{r^2 - a^2} \right\}.$$

For proof of this formula see Fleming's " Alternate-Current Transformer," Vol. I. Appendix, Note B. Second Edition.

To determine the induction which this force produces, wind upon the ring over the primary coil a small secondary coil consisting of ten or twenty turns of No. 36 S.W.G. silk-covered copper wire, interposing a layer of silk between the primary and secondary coils. This secondary coil must be connected with a ballistic galvanometer through a variable resistance. The primary coil must be connected with a battery through a reversing switch which will enable the primary current to be instantaneously reversed. These preparations being made, begin by standardizing the ballistic galvanometer with a standard condenser in the manner described in Elementary Form No. 5, and obtain the ballistic constant, or the number by which the sine of half the angle of throw of the coil or needle, when corrected by the logarithmic decrement factor, must be multiplied to obtain the quantity of electricity in micro-coulombs which passed through the galvanometer to produce that deflection. Then place an ammeter, carefully standardized (preferably one of Weston's ammeters), in series with the primary coil, and begin a series of observations. Close first the primary circuit, and observe the value of the primary current A on the ammeter in amperes. Then bring the ballistic galvanometer to rest, and connect it to the secondary coil. Suddenly reverse the primary current, and note the throw 2θ of the image of the spot of light reflected from the mirror of the ballistic galvanometer. One or two preliminary trials may be necessary in order to determine exactly the best resistance to have in the galvanometer circuit to obtain the greatest value of the throw of the coil or needle. Measure the total resistance of the secondary circuit, consisting of the galvanometer, secondary coil resistances, and leads. Let this altogether be R ohms. Then, if C is the ballistic constant of the galvanometer, we know that $C \sin \frac{\theta}{2} \left(1 + \frac{\lambda}{2}\right)$, where λ is the logarithmic decrement of the galvanometer, is the value in micro-coulombs of the quantity of electricity sent through the galvanometer by reversing A amperes in the primary coil. If the value of the induction density, or number of lines of force per square centimetre, in the iron core is called B, if S is the cross-section of the iron core, and if n is the number of turns of curve on the secondary coil, then B S n is the total induction through the secondary coil, and, as in Elementary Form No. 8,

$$\text{B S}\, n = \frac{1}{2}\, \text{R C} \sin \frac{\theta}{2} \left(1 + \frac{\lambda}{2}\right) 10^9 \times 10^{-7}.$$

Hence we can calculate the value of B for

$$B = \frac{\text{R C} \sin \frac{\theta}{2} \left(1 + \frac{\lambda}{2}\right) 10^9}{2\, \text{S}\, n\ 10^7},$$

where the factor 10^9 comes in to reduce the resistance R measured in ohms to absolute

C.G.S. units and the factor 10^{-7} to reduce the quantity measured in micro-coulombs to absolute C.G.S. measure. B is given then in C.G.S. measure. The factor 2 or $\frac{1}{2}$ comes in in the above equation because the current of A amperes is *reversed* in the experiment and not merely stopped. Hence we have the means of calculating the value of the magnetic force H in the place where the iron is, and also of measuring the value of the induction density B in it : and hence we can obtain the value of the permeability μ by taking the ratio of B to H.

If the sample of iron cannot be procured in the form of a ring we may proceed as follows : —Prepare as above two half rings of best soft iron of highest permeability attainable, and measure and wind them as above. Let the end faces of the half rings be perfectly plane, so as to fit together truly. Prepare the iron of which the permeability is to be taken in the form of two small cylinders about one centimetre in diameter and one centimetre in length, and let the end faces be truly plane and parallel. Wind secondary coils as above over these little cones. Arrange these small test pieces with coils between two half circular rings of soft iron of one centimetre in diameter in cross section, and press the whole together firmly. Then proceed to take the permeability as above described, using both the secondary coils as search coils in separate experiments. The primary coils on the two half rings must, of course, be joined in series. In calculating the magnetic force H, to which the test pieces are subjected, we may consider that it is nearly the same as would exist if the ring built up of the two half iron rings and two test pieces formed one single undivided ring. Hence, if l is the length of the test piece, and if l_1 is the mean length of each half iron ring, the whole mean length of the magnetic current, when put together, is $2 (l + l_1)$, and the mean radius of this compound ring must be considered to be equal to r_1 where $2 \pi \, r_1 = 2 \, (l + l_1)$, and this value of r_1 must be used instead of r in the above formula for H.

In order to secure good results the test pieces must have their end surfaces exceedingly true and flat, and must fit exactly against the true surfaces of the ends of the half ring magnets. The induction then flows uniformly through the test pieces without leakage, and the secondary coil wound on the test piece gives the value of this induction. The student should in this manner make measurements of the permeability of samples of iron and steel, and for fuller information on this subject he may refer to the following works :—

"Magnetic Induction in Iron and other Metals," Prof. J. A. Ewing, Chap. III., p. 59.

Article, "Magnetism," in "Encyclopædia Britannica," IXth Edition, Vol. XV., p. 256.

Mascart and Joubert, "Electricity and Magnetism," Vol. II., p. 637.

The Student should enter up his results in the form on the next page.

DETERMINATION OF MAGNETIC PERMEABILITY.

Mean diameter of iron ring =

Mean diameter of cross-section of iron ring =

Number of turns of primary coil =

Length of test piece =

Diameter of test piece =

Number of turns of secondary coil =

Resistance of whole galvanometer circuit =

Observation No...........	Current in primary coil in amperes.	Angular deflection of galvanometer needle on reversing primary current = θ.	Value of $\sin \frac{\theta}{2}\left(1 + \frac{\lambda}{2}\right)$	Calculated value of magnetising force = H.	Calculated value of magnetic induction, B.	Value of Permeability $\frac{B}{H} = \mu$.

ELECTRICAL LABORATORY NOTES AND FORMS.

Name   *Date*

The Standardization of a High Tension Voltmeter.

In these tests it is assumed that the Laboratory is supplied with alternating currents, which can be raised to a high pressure by transformers. The most convenient appliance for laboratory work is an alternator coupled direct to a continuous current motor, the motor being driven by current from secondary batteries. By suitable resistances it is then possible to drive at constant speed and to generate a steady alternating current at a pressure of, say, 100 volts. This can be raised by a small transformer to 1,000 to 3,000 volts. A series of divided resistances are required and also a potentiometer set.

The Student should carefully sketch out a diagram of the arrangement of apparatus required before proceeding to work.

Note.—In all tests with high voltage currents great care should be taken to avoid accidents by shock. India-rubber gloves should be used by the operators, and no circuit be handled when "alive."

The most commonly used high tension voltmeters employed for alternating current work are the electrostatic instruments. In these instruments there are a set of fixed plates or surfaces and a set of plates or surfaces which are movable round an axis or wire suspension. When a difference of potential is made between these fixed and movable plates there is an attractive force exerted between them which moves the movable plates against a restoring force due to gravity or to torsion. High tension electrostatic voltmeters of this class are very useful for alternating current measurements, because they are unaffected by the frequency of the alternations. In order to calibrate such a voltmeter, it should be fixed in position, well insulated, and then connected with the source of alternating electromotive force. The above-mentioned combination of motor and alternator is especially useful for such calibration work, because the value of the alternating potential is so easily changed and controlled by resistance inserted in the field of the continuous current motor. Suppose the voltmeter reads from 1,000 to 2,400 volts, we have then to provide a non-inductive resistance capable of being put in series across the high tension circuit and kept there for any length of time. The most convenient and safe resistances for this purpose are Dr. Fleming's non-inductive resistance cages of platinoid wire, which are made up in cages of 100 ohms or more. Twenty of these joined in series make a useful form of high tension resistance. From the ends of a section of one-twentieth of this resistance potential wires are led down to a low volt electrostatic voltmeter, preferably Lord Kelvin's horizontal pattern. The arrangement then is as follows:—The high tension voltmeter and high resistance are placed across the high tension mains, and a pair of potential wires are then joined to a small section of this resistance and brought to a low tension voltmeter. The low tension voltmeter must first have been calibrated carefully by the potentiometer.

In many forms of electrostatic voltmeter there is a small difference in the reading on a continuous current circuit, according to whether the needle of the electrostatic voltmeter is made *positive* or *negative*, and the difference between these readings may amount to a third or half a volt. Hence the low tension voltmeter must be calibrated with the potentiometer, both with the needle *positive* and the needle *negative*, and the mean value of these corrections taken as the correction for alternating currents. Thus, if the voltmeter reads 99·5 when 100 volts is applied to it by the potentiometer, needle being negative, and 99·8 when the needle is positive under the same conditions, we take 99·65 as the reading when 100 alternating volts are applied to it, and the scale error at this point is 0·35 for alternating pressure. The low pressure voltmeter being calibrated, we have then to measure very carefully the ratio in which the resistance is divided. This is done by measuring the whole resistance, and the resistance of the two sections.

Let the whole resistance be R ohms, and the resistance of the small section r ohms, then $\dfrac{r}{R}$ is the fractional division of the wire. The alternating pressure is then varied so as to put different pressures on the terminals of the high tension voltmeter. The corresponding reading of the low tension voltmeter is taken. If the high tension voltmeter reads a value, say, V volts by its scale, and if the *corrected* value of the low tension voltmeter at that instant is v volts, then the true value of the high tension pressure is $\dfrac{R}{r} v$ volts, and the error of the high tension voltmeter at that point on the scale is

$$V - \frac{R}{r} \cdot v.$$

There is one curious possible cause of error. The electrostatic voltmeters, although generally supposed to take no current, really have a small but definite capacity, and take a definite but small *alternating* current through them. Hence, if the value in ohms of the divided resistance is very large, say 40,000 or 50,000 ohms, the actual current flowing through this resistance may not be so very different from the capacity current of the electrostatic voltmeter used for the low readings, and this last may shunt a sensible fraction of the current. In other words, it can be shown that the ratio of the fall of volts down the whole, and down the small section of the resistance, is then no longer in the exact ratio of the whole to the small section of the resistance. Hence it is essential to have a resistance which does not take too small a current. In testing high tension voltmeters up to 2,000 volts, it is best to have a resistance of about 4,000 ohms, which will carry a current of 0·5 ampere, and to divide this resistance in the ratio of 20 : 1. The smaller section then has a resistance of 200 ohms, and the half ampere carried by it is very large compared with the capacity current of a low reading electrostatic voltmeter. Of course this source of error only comes in with alternating currents. Since the correction to be applied in the case when we are using a very high resistance involves a complicated function of the resistances and inductances of the sections of the wire as well as a knowledge of the capacity of the voltmeter, and since these last two quantities are difficult to measure accurately, the best plan is to avoid the necessity for correction altogether by employing a non-inductive divided resistance, taking a current of not less than 0·5 ampere when placed across the high tension circuit ; and this, together with the use of one of Lord Kelvin's multicellular electrostatic voltmeters to read the low tension volts, will eliminate any necessity for correction in most cases.

The Student should in this manner calibrate and check one or two commercial high tension voltmeters, and refer all the readings to the E.M.F. of a Clark cell.

The results should be entered up in the appended form.

Note.—H. T. is a common abbreviation for the words *high tension*, and similarly L. T. for the words *low tension*, these terms being applied broadly to pressures of about 100 volts and under and to pressures of 500 volts and upwards.

STANDARDIZATION OF A HIGH TENSION VOLTMETER.

Observation No.	Scale reading of H.T. voltmeter = V.	Value in ohms of whole resistance used = R.	Value in ohms of small section of resistance = r.	Corrected scale reading of L.T. voltmeter = v.	Error of H.T. voltmeter = $V - \frac{R}{r}v$.	Description of H.T. voltmeter used.

ELECTRICAL LABORATORY NOTES AND FORMS.

No. 29.—ADVANCED (No. 9).

Name   *Date*

The Examination of an Alternate=Current Ammeter.

The apparatus required for these tests is a carefully standardized ammeter which has been calibrated by continuous currents with the potentiometer in the manner described in the Advanced Form (No. 6). One of the Weston instruments or Kelvin ampere balances is a good one for this purpose. The alternating current ammeter to be checked is put in series with this standard, and is calibrated by continuous currents by reference to the standard direct-current ammeter.

The Student is recommended to sketch carefully the arrangement of the apparatus as set up.

An alternating current is one which reverses its direction at regular intervals, and passes through a uniform cycle of values. The number of complete cycles per second is called the *frequency* of the current. If a horizontal line is drawn, and divided into equal parts to represent equal small intervals of time—say, one-thousandth of a second—and if at each of these points an ordinate is erected representing in direction and magnitude the current in the conductor at that instant, the extremities of these ordinates delineate a curve which is called the *current curve*, and which has very different forms in different cases. In some cases it is nearly a simple sine curve ; in others, quite different. If the time line is divided into very numerous equal parts, and if ordinates are drawn at each of these divisions to meet the curve, *the square root of the average value of the square of the length of each ordinate* is called, for shortness, the *mean-square* value of the ordinate. This mean-square value (which is written $\sqrt{\text{mean}^2}$) is an important quantity. (For further information the Student should consult "The Alternate-Current Transformer," by Fleming, Vol. I., Chapter III.—Simple Periodic Currents, page 101, Second Edition.) If the curve is a simple sine curve the $\sqrt{\text{mean}^2}$ value of the ordinate can be shown to be equal to the maximum value divided by $\sqrt{2}$. But this rule does not hold good for forms of curve other than the simple sine curve. The measurement of alternating currents is made to depend either upon their heating power or upon the electro-dynamic attraction between different parts of the same circuit. If a current flows through a circuit, the rate of heat production depends on the square of the current strength at that instant. Hence, if the current is periodic, the mean rate of production of heat depends on the mean-square value of the current. If two parts of the same circuit are parallel to each other, and if one is fixed and the other movable, and if a current flows through them, the attraction or repulsion between these parts of the conductor depends on the square of the current strength in them. Hence, if the current is periodic, the mean force or stress between them will depend on the mean-square value of the current.

An alternating current of one ampere is thus defined : It is a periodic current such that its $\sqrt{\overline{\text{mean}}^2}$ value is unity, or one which will produce the same heating effect as a continuous current of one ampere would do when passed through the same conductor. The most useful type of instrument for measuring alternating currents is a dynamometer. Of these there are many kinds, such as the Siemens dynamometer, Lord Kelvin's ampere balances, and other similar instruments. In all of them, however, the current flows through a conductor, part of which is fixed and is called the stationary coil or coils, and part of which is movable and placed near the stationary part. The current flowing through the fixed and movable parts exerts a force drawing them together, and this stress is resisted either by the torsion of a wire or spring, or by gravity. If the time of free vibration of the movable part is very large in comparison with the duration of one complete cycle of the current, then the mean stress or force between these two parts is a measure of the mean-square value of the current if this should be periodic in value. The alternating current is measured in such an instrument by observing either the torsion or couple required to bring back the movable part to an assigned position with reference to the fixed coil, or else by the displacement of the movable part as observed on an arbitrary scale.

To calibrate a Siemens dynamometer, set it up in series with a standard calibrated direct current instrument, and send the same steady continuous current through them both. Observe the scale readings of the Siemens instrument for each particular value of the direct current. These same scale readings give the correspondingly valued alternating current. In the Siemens instrument the scale is usually a scale of equal divisions, 400 to the circumference. The number of degrees of torsion required to be given to the spring in order to bring back the movable coil to its normal position is indicated by a pointer moving over this scale. If the degrees of torsion required to bring the movable coil back to the zero position when one ampere is flowing through the coils is known—call it n—then the number of degrees of torsion required when A amperes is flowing through the coils is $n \, A^2$, and n is called the constant of the instrument.

In the case of instruments to be used for measuring alternating currents, large errors are likely to be introduced if any metallic parts are placed near the movable coil, by reason of the fact that eddy currents are set up in these parts and react upon the movable coil. Hence no metallic frames, cases or supports must be used, and this source of error must be guarded against. The Student should calibrate a few alternating current ammeters, using, if possible, currents of different frequencies. He will then find that many alternating current ammeters with iron in their coils do not give the same readings with different frequencies, and it will be seen that certain types of alternating current ammeter are only useful if calibrated with the current with which they are to be used.

Hence the above described process of calibrating with a direct current must be applied with caution. The best method to employ is to use a dynamometer or ampere balance having no iron in the coils at all. Then, in order to ascertain whether there is any sensible action produced by eddy currents set up in the framework of the instrument, pass the strongest alternating current possible through the movable coil, but nothing through the fixed coil. Note if the movable coil is at all displaced. If not, then calibrate this dynamometer by the ampere balance with direct currents. Then place it in series with the alternating current instrument to be standardized, and pass an alternating current through the two derived from the supply on which the alternating current instrument is to be used. Calibrate the ammeter by reference to the dynamometer. In testing an alternating current ammeter examine the following points :—

(i.) Ascertain if the instrument gives the same reading for alternating currents of different frequency,. but which have the same mean-square value as ascertained by a correct dynamometer. .

(ii.) If the instrument is a dynamometer, ascertain if there is any deflection of the movable coil when a current flows through it, but none through the fixed coil.

(iii.) If the instrument contains iron, note if the reading of the instrument is the same for the same current when reached by ascending from a smaller current as well as when descending from a larger current.

(iv.) Try if the instrument is affected by the presence of permanent magnets.

The Student should examine in this manner one or more forms of commercial ammeters for alternating currents.

EXAMINATION OF AN ALTERNATING AMMETER.

Observation No.	True value of the current through instrument.	Scale reading of the instrument.	Error of the instrument.	Frequency of the current.	Remarks.

ELECTRICAL LABORATORY NOTES AND FORMS.

No. 30.—ADVANCED (No. 10).

Name   *Date*

Delineation of Alternating Current Curves.

The apparatus required for these experiments is an alternator, on the shaft of which is a revolving contact-maker so arranged as to close a circuit at any assigned instant during the passage of the armature coils between the field magnet cores. Other apparatus required includes a set of 50 small secondary cells, an electrostatic voltmeter, and a condenser of about one-third microfarad capacity.

The Student is recommended to sketch carefully the arrangement of the apparatus as set up.

An alternator of any form can easily have a curve tracer applied to it, by means of which the form of the alternating current curves can be drawn. On one end of the shaft of the alternator (see Fig. 1) is fitted a gunmetal disk, which carries an ebonite disk about

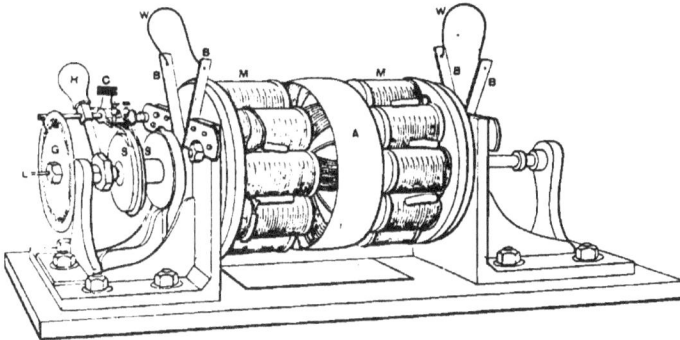

Fig. 1.—Alternator provided with Curve Tracer.

four inches in diameter and half an inch wide. The disk must be turned up on the shaft so as to be exactly centred. In the edge of the disk is let in a small slip of steel about one-sixteenth of an inch in width and as long as the disk is wide. Two small insulated springs, S S, are arranged on a rocking arm, H, which is centred on a fixed external pivot exactly in line with the centre of the armature shaft. These springs must be placed to press against the edge of the ebonite disk, and are then connected together electrically each time the steel slip passes underneath them both. The brushes, B B, can be rocked over so as to make contact at any angular position with reference to a fixed starting point, the angular displacement being recorded on a circular divided scale G. The arrangement of the spring holder and rocking arm will be best understood from the diagram in Fig. 1, in which is represented a small alternator provided with such a contact-maker.

Assuming that the laboratory is provided with this appliance, a number of interesting experiments can be carried out. A pair of insulated wires, W W, are connected to the two insulated contact springs, and another pair of insulated wires are connected to the terminals of the alternator and the four wires brought to the experimental table or laboratory. In the first place, let it be assumed that the alternator gives a pressure ($\sqrt{\text{mean}^2}$ value) of 100 volts.

To describe the electromotive force curve of the alternator, an electrostatic voltmeter, preferably one of Lord Kelvin's multicellular electrostatic voltmeters, has its terminals short-circuited by a well-insulated condenser of about one-third or one-quarter microfarad capacity, and the combination is then joined across the alternator terminals with the contact-maker interposed on one side. The effect of this arrangement is that, when the alternator is running, the contact-maker closes the circuit of the voltmeter at every revolution at an instant depending on the position of the rocking arm. The reading of the voltmeter is then the value of the instantaneous electromotive force of the alternator at the instant during the complete period corresponding to the position of the brushes. By moving the rocking arm over into various angular positions by the handle H the varying electromotive forces at different instants during the phase can be obtained. Since the electrostatic voltmeter does not read on its scale below a certain pressure of 50 or 60 volts, it is necessary to introduce a constant electromotive force in addition to that of the alternator to obtain a sufficient deflection for those values of the instantaneous electromotive force which lie below 50 or 60 volts. This is achieved by inserting in series with the voltmeter a variable number of small secondary cells, the electromotive force of which is separately measured on the voltmeter, and this value deducted from the total scale reading when the battery is so used. This blocks up the voltmeter to a false zero, and enables volts to be read by it down to zero. In this manner the value of the instantaneous electromotive force of the alternator can be observed and plotted down on a curve in terms of the angular intervals of one complete period, and such a curve is called an alternating current curve.

If it be desired to describe the current curve of the alternator, a non-inductive resistance, of such size as to carry the current comfortably without heating, is interposed in the external circuit of the alternator. The voltmeter, battery, and contact-maker are joined in series across this resistance. The curve of potential difference at the ends of this resistance is then the curve of current of the alternator. In taking the curve of terminal electromotive force of a transformer on the circuit of a high tension alternator it is necessary to divide up the potential and measure only a fraction of it. For this purpose a non-inductive resistance is joined across the high tension circuits. This resistance must be of such a form that it can be placed across the mains safely and carry a current which is not less than, say, half an ampere. Thus, if the circuit is a 2,000-volt circuit, this resistance should have a value of 4,000 or 5,000 ohms. A connection is made to a point of the resistance, which is one-twentieth of the whole, and potential wires are brought from this point and joined to the extremities of the voltmeter-battery and contact-breaker in series. In this way we can plot out a curve the ordinates of which are one-twentieth of the value of the potential across the whole main circuit. The reason for stating above that the divided resistance should carry a large fraction of an ampere is that, if the divided resistance is a very high resistance—say, 40,000 or 50,000 ohms—the current through it is not large compared with the capacity-current of the electrostatic voltmeter ; and under these conditions the voltmeter reading would not be the same fraction of the whole potential difference between the mains that the resistance of the small section of the resistance is to the resistance of the whole. The proof of this fact is somewhat long, but the general nature of the effect can be understood by bearing in mind that electrostatic voltmeters have a small but measurable capacity, and that, although they would not pass any continuous current, they do permit the passage of an alternating current, which is called the capacity-current of the voltmeter. If the voltmeter is joined across a section of a very high resistance, it sensibly shunts some current and reduces the effective resistance of this section. The curve of primary terminal potential difference of a transformer can, however, be described in the above manner. The curve of primary current can be also obtained by putting in series with the transformer a resistance of such magnitude as to afford a drop of about 100 or 150 volts to measure at a maximum. Having described the curves of electromotive force and current, the curve of induction of the transformer may be obtained by integrating the curve of primary potential difference. A set of curves so drawn for a 10-kilowatt transformer are shown in Fig. 2. For further infor-

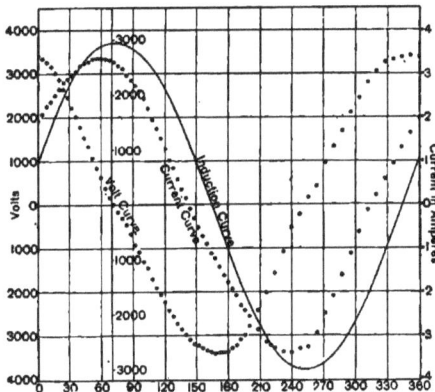

Fig. 2.—Curves of Current and Electromotive Force of a 10·H.P. Transformer taken off a Kapp Alternator.

mation as to the manner of obtaining the curve of induction from the curve of electromotive force, and for fuller details of the method of instantaneous contacts, the Student is referred to " The Alternate-Current Transformer," by Fleming, Vol. II., Chapter IV. For a description of a method of measuring these transformer curves when the alternator is not accessible, the Student is referred to a paper in *The Electrician*, 1895, Vol. XXXIV., pages 460, 507.

The Student should make a complete study of an alternator and a transformer by this method, and record the observations in the appended Forms, plotting down the curves on squared paper.

DELINEATION OF ALTERNATING CURRENT CURVES

Electromotive Force Curve.

Ratio of divided resistance used

Observation No.	Scale reading of contact-maker.	Voltmeter reading.	E.M.F. of battery in series with voltmeter.	Corrected instantaneous E.M.F. of alternator.	Phase angle in degrees.

Current Curve.

Resistance used in circuit = ohms.

Observation No.	Scale reading of contact-maker.	Voltmeter reading.	E.M.F. of battery in series with voltmeter.	Corrected instantaneous E.M.F. at terminals of resistance.	Corrected instantaneous value of the current.	Phase angle in degrees.

ELECTRICAL LABORATORY NOTES AND FORMS.

Name   *Date*

The Efficiency Test of a Transformer.

The apparatus required for these experiments is an alternating current transformer, and a means of supplying it with alternating current at a pressure of 1,000 or 2,000 volts. The load on the secondary circuit of the transformer must consist of incandescent lamps. The instruments required are an ammeter and voltmeter for reading the primary current and pressure, and other similar instruments for reading the secondary current and pressure, and also a non-inductive wattmeter. Rheostats and resistances must be provided for regulating the primary pressure.

The Student is recommended to sketch out very carefully the arrangement of circuits, and to examine them well before beginning the test. In these and all other high-tension experiments, it is an ESSENTIAL PRECAUTION to wear a pair of indiarubber gloves and to stand on an indiarubber mat, to avoid risk of accident.

An alternate current transformer consists of two circuits wound over an iron core, called the primary and secondary circuits. One of these, generally called the primary, is employed to transmit a high pressure alternating current, and the other, called the secondary, has then produced in it another current at a lower pressure, called the secondary current. When the transformer is at work there is a certain difference of pressure between the primary terminals, which varies in a periodic manner, and if a high-tension electrostatic voltmeter is joined across these terminals the voltmeter will measure the $\sqrt{\text{mean}^2}$ value of this alternating pressure, and this is called the primary potential difference (P.P.D.). In the same way an electrostatic voltmeter placed across the secondary terminals will measure the $\sqrt{\text{mean}^2}$ value of the secondary potential difference (S.P.D.). This symbol $\sqrt{\text{mean}^2}$ stands for the long phrase " the square root of the mean of the squares of all the instantaneous values of the varying quantity (current or electromotive force) taken at numerous equidistant intervals during the complete period."

If a Siemens dynamometer is inserted in series with the primary circuit of the transformer, we can in like manner measure the $\sqrt{\text{mean}^2}$ value of the primary current. The transformer to be tested should have a load of lamps or other non-inductive resistance arranged on its secondary circuit, and should have an electrostatic voltmeter, carefully calibrated, placed across its secondary terminals and one across its primary terminals; also a Siemens dynamometer or current balance in series with its primary and secondary circuits. When the load on the secondary circuit is practically a non-inductive load, the product of the $\sqrt{\text{mean}^2}$ value of the secondary terminal potential difference in volts

(S.P.D.), and the $\sqrt{\text{mean}^2}$ value of the secondary current (S.C.), as given by the instruments, gives us truly the mean power given out in watts on the secondary external circuit. The primary circuit of the transformer is, however, an inductive circuit, and hence the product of the $\sqrt{\text{mean}^2}$ values of the primary potential difference (P.P.D.) and primary current (P.C.) does not always, or at least when the transformer is lightly loaded, give us the true mean power given to the transformer. The product of P.P.D. and P.C. is called the *apparent power* or *apparent watts* given to the transformer. The real mean power or true watts can only be ascertained by a proper wattmeter constructed as follows :--

The series coil of the wattmeter must be of wire sufficiently large to carry the full primary current of the transformer, and, if that is considerable, it is best made of stranded wire. The shunt coil of the wattmeter must not have more than five or six turns of wire on it, and this should be thick wire capable of carrying from ten to twenty amperes. The wattmeter should be constructed without any metal parts near the movable coil. The movable coil is to be connected to the secondary circuit of a small transformer which supplies it through a bank of lamps or resistance with current at low pressure. This auxiliary transformer has its primary terminals connected to the primary mains of the transformer under test.

In the diagram (Fig. 1) the wattmeter is represented by W, S being the series coil, and Sh the shunt coil. The auxiliary transformer is represented by T_1. Its secondary circuit includes a lamp or lamps L, and the shunt coil Sh. The transformer to be tested is represented by T, and the keys k_1 and k_2 close the primary circuit of this transformer, or else the non-inductive resistance in parallel with it.

The wattmeter may be preferably one of the Siemens form with spiral torsion spring and graduated dial. To standardize this wattmeter it is necessary to provide a non-inductive resistance, which is of a form suitable for putting across the high-tension mains, and it will then absorb a power which must be comparable with the power to be measured. A special form of resistance for this purpose has been designed by Dr. Fleming. This non-inductive resistance must be so arranged that, by means of keys k_2 or k_1, the resistance or the primary circuit of the transformer can be placed in connection with the wattmeter, the arrangements of circuits being as in the diagram.

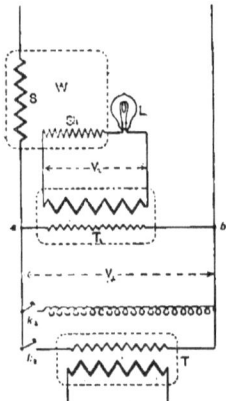

Fig. 1.

Having connected up the apparatus properly, a series of observations should be taken of the power taken up by the transformer and that taken up by the non-inductive resistance, as follows. The non-inductive resistance should be divided into two sections, and one of these sections should be of such current-carrying capacity that it is not sensibly heated by the current which will flow through the whole resistance when placed across the high-tension circuit. This small section of the resistance should have such a magnitude that under these circumstances the fall of volts down it is about 100 volts ($\sqrt{\text{mean}^2}$ value). Connect the whole resistance across the primary mains and regulate the primary pressure until it has the standard value at which the transformer is to be tested. Let the value of the small section of this resistance be r ohms and the value of the whole resistance R ohms. Let v be the fall of volts down the small section and V that down the whole section. Then $V\frac{v}{r}$ represents in watts the mean value of the power taken up in the non-inductive resistance when a terminal $\sqrt{\overline{\text{mean}^2}}$ pressure of V volts is applied to it. This preliminary test should be carefully made. The non-inductive resistance is then applied to the wattmeter, and the reading of the wattmeter taken when the primary pressure has the standard value V. Suppose the scale reading of the wattmeter is w, then the true watts corresponding to w is $\frac{V}{r}v$ and the value of one scale division of the wattmeter is $\frac{V\,v}{w\,r}$ watts. The wattmeter is then switched over on to the primary circuit of the transformer, and the wattmeter reading taken. Let it be W scale divisions ; then the mean power being taken on the primary of the transformer is equal to $\frac{W\,V\,v}{w\,r}$ watts. Let the primary current of the transformer at this instant be A_1 ($\sqrt{\overline{\text{mean}^2}}$ value). The apparent watts taken up by the transformer is A V. The ratio of the true watts to the apparent watts is called the *power factor* of the transformer.

At the same time the $\sqrt{\overline{\text{mean}^2}}$ value of the secondary current should be taken, and also that of the secondary potential difference at the terminals. Let these values be A^1 and V^1 amperes and volts respectively. Then $A^1 V^1$ is the mean power in watts delivered up by the transformer. The efficiency of the transformer is the ratio of power coming out to power going in, or the efficiency e is given by the fraction

$$e = \frac{A^1\,V^1}{\dfrac{W\,V\,v}{w\,r}} = \frac{A^1\,V^1\,w\,r}{W\,V\,v}.$$

This efficiency should be taken at various fractions of full output. These results should be plotted down in the form of two curves, one showing the increasing efficiency in terms of the secondary load, and the other showing the total loss in the transformer in terms of the same. The observer should then measure the resistance of the primary and secondary circuits of the transformer and calculate the copper losses at all loads, and should take a particular observation to determine the loss in the transformer when the secondary circuit is open. This last is called the iron loss or core loss of the transformer, and it is found that the total loss in the transformer at any load can be calculated by taking the sum of the iron loss and the total copper losses at that load, allowance being made for any rise in temperature in the transformer circuits.

The Student should make in this way a careful study of one or two transformers, and enter the results in the following tables.

EFFICIENCY TEST OF A TRANSFORMER.

Transformer No.   by
Nominal power   kilowatts.
Resistance of primary circuit =   ohms at   °C.
Resistance of secondary circuit =   ohms at   °C.
Transformation ratio
Standardizing resistance =   ohms.
Power taken up at   volts in resistance =   watts.
Wattmeter calibration =

Observation No.	Primary volts.	Primary current.	Secondary volts.	Secondary current.	Wattmeter reading.	Watts into primary.	Watts out of secondary.	Efficiency.

Observation No.	Primary current.	Secondary current.	Copper loss in primary.	Copper loss in secondary.	Iron core loss at no load.	Total loss in transformer.	Fraction of full load.

ELECTRICAL LABORATORY NOTES AND FORMS.

No. 32.—ADVANCED (No. 12).

Name   *Date...*

The Efficiency Test of an Alternator.

The plant required for these tests is an alternator and a continuous current motor. Also the necessary instruments for reading the currents and pressures of both circuits. A non-inductive resistance capable of absorbing the electrical output of the alternator is required. If the alternator and motor can be put down upon the same bedplate, with shafts in line and connected by a flexible coupling, this arrangement is the best. If not, the pulleys of the two machines must be connected by a fairly long flexible belt, with good adhesion.

The Student should sketch the arrangement of the plant and circuits.

The practical problem is very frequently presented to the experimentalist of testing an alternator when only that one machine is available. The most direct method of attacking the problem is to obtain a continuous current motor of sufficient power to drive the alternator, and to determine first the efficiency of that motor at the speed and the loads at which it will have to run when used to drive the alternator. This is easily done by the brake method, described in Elementary Form No. 18, or, if two equal motors are available, by the Hopkinson method, described in an Advanced Form (No. 20). The efficiency of the motor being determined, we know exactly how much power is produced on the pulley of the motor when any definite number of watts is being sent into the motor. If, then, the motor is connected to the alternator, we have a means of knowing approximately what is the power given to the pulley of the alternator. If this connection is a direct one—that is, if the shaft of the alternator is coupled directly to the shaft of the motor, then the power given out by the motor must be the power given to the alternator ; but if the connection has to be made by a belt, then we have to take account of the work lost in bending the belt and in that due to slip of belt.

This may be done approximately by mounting the pulley of the alternator on a shaft running in bearings very freely and kept well oiled, and stretching the belt over this pulley and the pulley of the alternator with about the same degree of tightness which is used in the actual experiments. This shaft should have a large iron disk keyed to it and a rope brake applied to it. We determine then the power given out on this freely moving shaft when a certain power in watts is given to the motor, and when the shaft is running at the proper alternator speed. In this estimate, therefore, we have included whatever power is required to bend the belt and that lost by the slip of the belt. The pulley is then replaced on the alternator, and the alternator circuits connected up through an ammeter with the non-inductive resistance which is to take up the power. Across the poles of the alternator is placed a voltmeter. Both these instruments must be calibrated for alternating currents of the frequency to be employed.

A continuous-current ammeter is also in the feeding circuit of the motor, and a voltmeter across its terminals ; both these instruments must previously have been calibrated. The combination having been set at work, we note the current A_1 in amperes going into the motor. This must include that going to excite the fields. We observe the volts V_1 across the terminals of the motor. The current A_2 coming out of the alternator must be taken, and also the volts V_2 across its terminals. The above-described experiments will enable an answer to be obtained to the question, How much power in watts is given to the pulley of the alternator when $A_1 V_1$ watts are given to the motor ? The product $A_1 V_1$ represents in watts the power given to the driving motor. Suppose we have found by the above-described experiments with the brake that this yields a power P watts on the pulley of the alternator. Then if the alternator is giving out a current A_2 at a pressure V_2, it is giving out $A_2 V_2$ watts, and the efficiency of the alternator is therefore given by $\dfrac{A_2 V_2}{P}$ expressed as a percentage. This efficiency should be taken for various loads in the alternator, and a curve should be plotted in which the horizontal distances represent outputs of the alternator, and the vertical ones efficiency of the alternator. At the same time we have to take into account the power required to excite the field-magnets of the alternator. To do this we must obtain the field current α in amperes and the volts v at the terminals of the field circuit, and add the product $a v$, or watts required to excite the field, to the power given to the pulley of the alternator (that is to P) before calculating the efficiency. Hence the correct efficiency of the alternator will be given by the fraction $\dfrac{A_2 V_2}{P + a v}$ expressed as a percentage.

The Student should in this manner test a small alternator and obtain the efficiency at various loads and speeds, and enter up the results in the appended form.

On the subject of testing alternating current machines, the Student is referred to a very excellent Paper, by Mr. W. M. Mordey, in the "Proceedings of the Institution of Electrical Engineers," Vol. XXII., 1893, p. 116. In this Paper it is shown that large alternators with fixed armatures may be tested by using one part of the armature as a motor, against another part as a dynamo. Various modes of testing large alternators are given.

EFFICIENCY TEST OF AN ALTERNATOR.

Alternator No.　　　　by
Armature resistance (warm) =
Field resistance =
Number of bobbins on armature =
Number of poles on fields =
Frequency =
Normal speed =　　　　. Full output =　　　watts.
Exciting current = a =
Exciting volts = v =

Observation No.	Current into motor = A₁.	Volts on terminals of motor = V₁.	Current out of alternator = A₊	Volts on terminals of alternator = Vₛ.	Speed in revolutions per minute = N.	Efficiency of alternator = e.

EFFICIENCY TEST OF AN ALTERNATOR.

Observation No.	Current into motor = A_1.	Volts on terminals of motor = V_1.	Current out of alternator = A_3.	Volts on terminals of alternator = V_2.	Speed in revolutions per minute = N.	Efficiency of alternator = e.

ELECTRICAL LABORATORY NOTES AND FORMS.

Name   *Date*

The Photometric Examination of an Arc Lamp.

The apparatus needed for these experiments is an arc lamp with very uniform and regular feed, and in the first instance had better be a continuous current arc lamp. This lamp must be suspended from the ceiling so that it can be raised or lowered and moved horizontally. A plane mirror must be provided, fixed to a horizontal rod or tube at an angle of 45 degrees, this horizontal tube being capable of revolution round its axis. The remaining apparatus comprises a photometric bench, photometer disc, standard incandescent lamp, and measuring instruments.

The Student is recommended to sketch carefully the arrangement of the apparatus.

If an arc lamp is set up and worked with continuous current, the positive or crater carbon being uppermost, a very little examination shows that the light is not uniformly distributed. A complete photometric study of an arc lamp for practical purposes involves several investigations. The distribution of light has to be examined. The mean spherical candle-power and corresponding electrical power absorbed has to be determined and the action of the regulating mechanism tested. The first question involved is that of the standard with which the light is to be compared. A slight experience of comparing the light of the arc with that of a gas flame or normal incandescent lamp shows that the difference in colour between the two lights is very marked, and that, when using a Bunsen disc, the unpractised eye is very perplexed to establish any comparison between them at all. It has usually been the custom to evade this difficulty by the use of red or green glass screens, and to measure what is popularly called the "red candles" or "green candles" of the arc. This, however, is practically a useless performance. The best standard to select for comparison is a good incandescent lamp, which is used at about 2 or 2½ watts per candle, and which under these conditions gives a light more suitable for a comparison standard than that of a glow lamp burning at 3 or 4 watts per candle. The very rapid blackening of the standard lamp can be minimised by employing a very large glass, of such size that the filament is four or five inches away from the glass envelope on all sides. This over-burnt incandescent lamp is kept constantly standardized by reference to a normal standard incandescent lamp or gas-flame standard. The best means of comparison is by a star disc or by Trotter's wedge photometer. In this last form of photometer two pieces of perfectly white card or metal, painted dead white, are set up on the photometer bench, their planes being vertical and meeting at an angle of about 80 degrees or 90 degrees. The vertical angle or meeting edge is pointed in one direction along the photometer bench. In that surface nearest the observer are cut one or more slits or holes. If two lights are placed one on either side of the wedge, it will be easily seen that an observer looking at the wedge will see the front or outer surface of the front card illuminated from one side, say the left-hand light. Looking through the holes in this card, he will see the inner surface of the other card illuminated by the right-hand light. The wedge can so be moved that these surfaces are equally illuminated, and then the hole will hardly be seen at all.

The process of obtaining this balance of illuminations is facilitated by swinging the wedge from side to side in slowly diminishing arcs. Whatever form of photometer is adopted, the process of photometry consists in effecting a balance or equality in brightness of two white surfaces, one of which is illuminated by the standard light and the other by the light under test. And in making this estimate, the eye has to discriminate the equality in brightness apart altogether from any outstanding difference in colour. The arrangement being thus made for the photometry of the arc, the apparatus should be set up as follows :—

The arc lamp should be suspended from the ceiling and be so capable of being moved that the arc can be traversed over a half-circle in a plane perpendicular to the direction in which it is to be photometered. The inclined mirror is set on its axis at 45 degrees at the centre of this circle, and provided with means for measuring the angles through which that axis is rotated ; the axis of the mirror must be in the direction of the central axis of the photometer. The mirror is so set that it can catch the rays sent out from the arc at any angle to the horizon and reflect them along the axes of the photometer, and the angle of reflection from the mirror always remains the same. By traversing the arc round a circle whose centre lies on the axis of the mirror, we can always keep the arc at the same distance from the mirror and reflect the ray at the same angle, and yet send the rays coming from the arc at any angle to the horizon along the horizontal axis of the photometer. The mirror must have its coefficient of reflection at 45 degrees determined as in Elementary Form No. 16.

The ray reflected horizontally from the mirror must then be photometered against the glow-lamp standard, care being taken to protect the photometer wedge from receiving any rays from the arc lamp except those which have been reflected from the mirror at a definite angle. By raising and moving the arc and shifting the mirror, the observer can determine the candle-power of the arc in various directions and angles to the horizontal. The distance of the arc from the photometer-wedge and the distance of the glow lamp standard from the same must be carefully measured. If I is the intensity or candle-power of the ray proceeding from the arc in any direction, and if a is the percentage of light reflected from the mirror, then $I\dfrac{a}{100}$ is the candle-power of the ray after reflection. If i is the candle-power of the glow-lamp, and if D and d are the distances of the arc and glow-lamp respectively from the wedge, then

$$\frac{I\dfrac{a}{100}}{D^2} = \frac{i}{d^2},$$

or

$$I = i\frac{D^2}{d^2}\frac{100}{a}.$$

Whilst one observer is taking the candle-power of the arc, another observer must take the value of the power taken up in the arc. This is done by placing an ammeter in series with the arc and connecting a voltmeter between the two carbons.

Small clips are provided which connect the voltmeter to the carbons about an inch or two above and below the arc. The ammeter should be so inserted in the circuit that it measures the current going though the arc above and takes no account of that passed by the shunt coil of the arc-lamp mechanism. The intensity of the rays sent off in different directions should be measured for every 10 degrees above and below the horizontal line through the arc, and the results then set out in a photometric diagram of the arc. A centre is taken at A (see Fig. 1) to represent the arc and a semicircle, B D C, drawn round it with radii spaced 10 degrees apart. On these radii are set out distances to represent to any scale the illuminating power in that direction in terms of the standard, and a curve, A P N, drawn passing through all these points. This curve

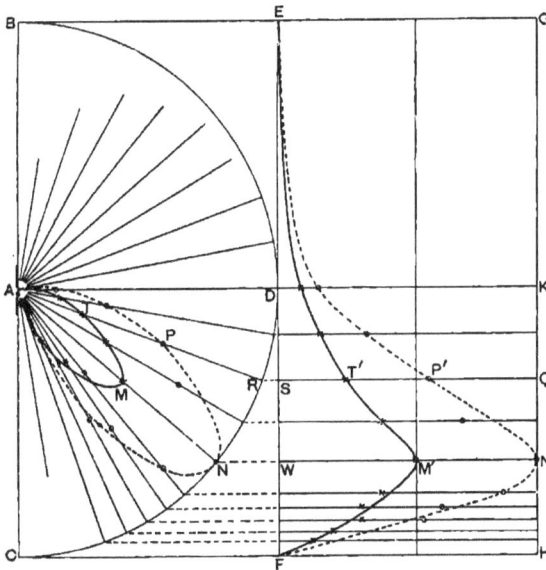

will then represent the candle-power of this arc in different directions. The mean spherical candle-power can be obtained from it geometrically as follows :—

Draw a vertical line, E F, which is a tangent to the semicircle at D, and which is therefore parallel to the vertical line through the arc. From the ends of the radii of the semicircle drawn at intervals of 10 degrees draw horizontal lines; and, starting from the vertical line, set off on these horizontal lines distances equal to the

Fig. 1.

illuminating power of the arc in those different directions. The extremities of these lines define another curve, as shown in the diagram in Fig. 1. Thus the ordinate S P' of the curve E P' N' F is equal to the radius A P of the photometric curve A P N.

The area included by this curve E P' N' F, and the vertical line E F, must be taken with the planimeter and compared with the area of the circumscribing rectangle E G H F. The mean spherical candle-power of the arc lamp is given by the product of its maximum candle-power in candles, represented by the lines A N or W N', and the ratio of the areas included by the above projected curve and its circumscribing rectangle.

THE PHOTOMETRIC EXAMINATION OF AN ARC LAMP.

Arc lamp tested —

Observation No.	C.P. of standard lamp.	Distance of standard from disk.	Distance of arc lamp from disk.	C.P. of arc lamp.	Angle to horizontal at which ray is being sent out.	Current through arc in amperes.	Potential difference of carbons in volts.	Power in watts spent in arc.

ELECTRICAL LABORATORY NOTES AND FORMS.

No. 34.—ADVANCED (No. 14).

Name...   *Date*...

The Measurement of Insulation and High Resistance.

The apparatus required for these tests is a sensitive movable coil galvanometer, some resistance boxes, and a megohm standard. A set of 200 small well-insulated secondary cells must be provided, which are preferably small Lithanode cells. The galvanometer, battery, resistance boxes, and all connections and keys must be very highly insulated. A long length, about half-a-mile, of insulated cable should be provided for test, and this should be placed in a tank of water with its ends out.

The Student should sketch the arrangement of all the connections.

Although many methods of testing high resistance have been proposed and are to be found described in electrical text books, there is only one method which yields results which are perfectly satisfactory in practice, and that method is the method of testing insulation resistance directly by the current sent through the insulation by a known electromotive force.

Suppose, in the first place, a test has to be made of the insulation resistance in megohms per mile of an insulated cable. The first step is to measure the length of the cable. If this is not already done and certified, the only way is to weigh the whole coil of cable and then weigh a known length, say fifty yards, of the cable, and deduce the whole length of the cable from the ratio of these weights. The ends of the cable must then be trimmed by removing, for a length of about one yard from each end, the cotton or hemp twist, tarred tape, or protective layer, but leaving the indiarubber or gutta percha insulation. These ends must then be carefully dried. The coil, with the exception of these ends, is then immersed in water at 70°F. in a tank and left there for 24 hours. The free ends must be tied up by silk strings so that they may not get wet. The copper strand is exposed for an inch at each end, and the two ends of the copper conductor of the cable may be twisted together.

The galvanometer must then be set up in a steady place, a divided scale placed at the proper distance, and a very sharp image of an incandescent lamp filament focussed upon it.

The battery is then set up and well insulated on slips of ebonite, and a highly insulated key placed in series with it. One end of the battery is then joined to the tank and the other end to the galvanometer. The other end of the galvanometer is joined to the copper of the cable. A key should be inserted *between the battery and tank.* Begin the experiment by connecting the megohm resistance in place of the cable and tank, that is, join the galvanometer, megohm resistance, and cells in series, and vary the number of cells until a deflection of the galvanometer is obtained which can be read to one per cent. Let the number of cells so used be n. Measure the voltage of these cells with an electrostatic voltmeter, and let it be V volts. Then, if the deflection of the galvanometer was d divisions, a current of $\frac{V}{10^6}$ amperes makes a deflection of d divisions of scale when it passes through the galvanometer. Next place the galvanometer, battery and cable as above in series, and, in the first place, use only one or two cells of the battery to make sure that the insulation is not defective. If it proves to be sound, gradually increase the pressure until a deflection of the galvanometer is obtained, either about as great as that given with the megohm or as great as the whole number of cells at disposal will allow. To be satisfactory, the number of cells should be not less finally than about 200, so as to give a pressure of 400 volts. If the deflection of the galvanometer is then d_1 divisions with V_1 volts acting through the insulation of the cable, and if this insulation is R megohms, we have $R = \dfrac{V_1 d}{V d_1}$.

Test the galvanometer and battery for leakage by removing the connection between the battery and the tank, when no galvanometer deflection should be found. It is convenient to short-circuit the galvanometer terminals during the first closing of the battery circuit to avoid the sudden and large ballistic deflection of the galvanometer which takes place at first contact owing to the capacity of the cable. When, however, the cable is charged, it will be found generally that the galvanometer deflection decreases from moment to moment as the time of electrification increases. This is caused by an *increase* in the dielectric resistance which takes place from moment to moment. It is generally usual to record the insulation resistance after one minute's electrification. With some kinds of dielectric it will be found that the insulation resistance goes on increasing for a very long time, and the student should note this and plot a curve showing the time-increase of dielectric insulation. The insulation resistance also decreases generally as the temperature rises, and hence it must be recorded at a known temperature, which is usually 70°F. Useful experiments can be made in the laboratory with about half-a-mile of insulated cable in a tank. The results of measurement are generally reduced to megohms per mile at 70°F. after one minute's electrification with a pressure of 400 or 600 volts. In testing house-wiring insulation, it is useless to use less than 100 volts pressure in testing. In such tests the circuit switches of the house should all be closed, so as to connect together all the wiring, and the lamps be removed. The insulation resistance of both positive and negative sides should be taken separately.

Very various rules have been laid down by different authorities as the standard insulation resistance which should be reached by good house wiring for electric lighting purposes. The Phœnix Fire Office requires a minimum insulation resistance of 12·5 megohms per lamp for continuous currents, and double this for alternating currents.

The Institution of Electrical Engineers recommend leakage not to exceed one five-thousandth part of the total working current. Various electric lighting companies adopt standards of from 10 to 100 megohms per lamp as the insulation resistance required in a building. The state of the atmosphere has, however, a great deal to do with the insulation measurement of a building completely wired and with all fittings on, because no inconsiderable portion of the leakage is not true insulation conduction in the cables, but surface leakage over the porcelain, slate, fibre, or wooden bases of sockets, fuzes, switches, cutouts, &c. The minimum insulation allowable in the cables and wires themselves for 100-volt work is about 300 megohms per mile after 24 hours soaking in water. For high-tension work at 2,000 volts the insulation resistance of cables should be 4,000 or 5,000 megohms per mile at least. For this latter class of work too much stress should not be laid on mere insulation resistance. The only satisfactory test is the actual breaking-down pressure in volts.

THE MEASUREMENT OF INSULATION RESISTANCE.

Length of cable or circuit tested =
Time of soakage in water =
All tests made after one minute's electrification.

Observation No.	Galvanometer deflection through standard resistance.	Battery E.M.F. in volts.	Resistance used to standardize.	Galvanometer deflection through insulation under test.	Battery E.M.F. in volts.	Calculated insulation resistance.	Circuit or cable tested.

THE MEASUREMENT OF INSULATION RESISTANCE.

Observation No.	Galvanometer deflection through standard resistance.	Battery E.M.F. in volts.	Resistance used to standardize.	Galvanometer deflection through insulation under test.	Battery E.M.F. in volts.	Calculated insulation resistance.	Circuit or cable tested.

ELECTRICAL LABORATORY NOTES AND FORMS.

No. 35.—ADVANCED (No. 15).

Name   · Date

The Complete Efficiency Test of a Secondary Battery.

The apparatus required for these tests is a potentiometer set and galvanometer, and also some low resistance standards capable of carrying the full discharge current of the battery. A bank of incandescent lamps as an absorbing resistance must be at hand, on which to run down the battery. An ammeter, voltmeter, and carbon resistance are also necessary.

The Student should sketch the arrangement of the apparatus and circuits.

The Student will be assumed to have a general acquaintance with the chemical changes taking place in a secondary battery during charge and discharge. When a secondary battery is charged during a certain time, a current is put into it at varying or constant potential. At any instant a certain current in amperes is being supplied to the battery, which will generally vary from instant to instant. If at any instant the value of the current flowing into the battery is taken by an ammeter, and if also the difference of potential is taken at the same instant by a voltmeter, the product of this current and potential difference in amperes and volts gives us the value of the power in watts being supplied to the battery at that instant. Supposing, then, that a straight line is taken, on which we mark off a length representing the time of charging, and at the proper intervals erect perpendiculars to represent to any scale the ampere-current flowing into the battery, and also the watt-power put into the battery, and join the tips of these perpendiculars by a curve, the area included between the time base line, the extreme perpendiculars, and the curves estimated in square units, representing an *ampere-hour* or *watt-hour*, will give us the total quantity and total energy respectively put into the battery. Then let the battery be discharged through a resistance, and the current flowing out of the battery and the potential difference at its terminals be continually noted and recorded. We can, on the same diagram, draw curves representing the discharge of the battery—in amperes and watts. The ratio between the ampere-hours taken out of the battery and the ampere-hours put into the battery is the *ampere-hour efficiency* (A.H.E.) of the battery, and similarly the ratio between the watt-hours taken out of the battery and the watt-hours put into the battery is the *watt-hour efficiency* (W.H.E.) of the battery. The A.H.E. will vary with the time the battery has been standing, and with the rate of discharge. The greater the mean current taken out of the battery, the less, generally speaking, will be the A.H.E. The W.H.E. varies also with the rate of charge and discharge. In order to charge the battery, the working or external E.M.F. must exceed that of the battery, which is two volts per cell, and it is usual to employ 2·5 volts per cell to charge. Hence, the W.H.E. does not generally exceed 80 per cent., and is often much less.

The Student should experiment upon a single cell or pair of cells. The weight and total exposed surface of the plates, positive and negative, separately and together, should be taken. The weight of the glass or other box and acid should also be taken. The cell should then be completely discharged and the battery joined up as follows :—

An ammeter and a carbon resistance should be placed in series with the cell, and a voltmeter across its terminals. The process of charging should then be begun and continued steadily until the bubbles of gas begin to come off freely, when it should be stopped. The cell should then be discharged through the carbon resistance and the volts and amperes of discharge observed. The charge and discharge curves should then be drawn, and the ampere-hours and watt-hours put in and taken out, reckoned up, and the efficiencies obtained. This should be done for several different rates of discharge, keeping the current constant by the adjustable carbon resistance. The total ampere-hours of charge should then be reckoned out per kilogramme or per pound of plates, taking both positive and negative plates together, and the capacity per square foot or square decimeter should also be taken, reckoning positive plate surface only. A curve should be drawn representing the varying decrease of ampere-hour capacity with increasing rate of discharge, and the same for the watt-hour efficiencies. In the case of a large cell or cells, the potentiometer may be used to measure the current and volts in charge and discharge, the battery charging or discharging current being taken through a resistance strip.

The weight of the plates should be taken again at the end of all the experiments, to ascertain how much they have lost by disintegration. Note should be taken of any buckling or bending of the plates. During the process of discharge the circuit should be opened at intervals, and the open circuit volts of the cell measured. This will enable the observer to calculate the internal resistance of the cell (*see* Elementary Form No. 12) corresponding to any given rate of discharge. The discharge should be considered to be over for practical purposes when the open circuit potential difference of the cell falls below 1·9 volts. All discharge beyond that point is useless for working incandescent lamps in parallel. Hence the ampere-hour and watt-hour discharge must be reckoned as complete when this point is reached.

The Student should enter up his results on the appended Forms and carefully set out all the results in the form of curves.

The Student is referred for much useful information on the process of charge and discharge of a secondary battery to a Paper by Prof. Ayrton in the " Proceedings of the Institution of Electrical Engineers," 1890, Vol. XIX., p. 539, entitled " The Working Efficiency of Secondary Cells," and to the important discussion which took place on this Paper.

TEST OF A SECONDARY BATTERY.

Cell or battery tested   =
Total weight of plates   =
Weight of box and acid =
Total surface of positive plates

CHARGE.			DISCHARGE.			
Ampere current into cell.	P.D. at terminals in volts.	Time.	Ampere- current out of cell.	P.D. at terminals in volts.	Time.	Internal resistance of cell.

TEST OF A SECONDARY BATTERY.

Observation No.	CHARGE.		DISCHARGE.		Efficiency.	
	Ampere-hours put in.	Watt-hours put in.	Ampere-hours taken out.	Watt-hours taken out.	A.H.E.	W.H.E.

ELECTRICAL LABORATORY NOTES AND FORMS.

No. 36.—ADVANCED (No. 16).

Name   *Date*

The Calibration of Electric Meters.

The apparatus needed for these tests is an ammeter and voltmeter, which should previously have been standardized with the potentiometer, and a variety of electric meters. The supply of current to the laboratory, in order to obtain perfect steadiness of current, must be from secondary batteries.

The Student should sketch carefully the arrangement of the apparatus as set up.

Electric meters may be broadly divided into ampere-hour meters and watt-hour meters. The former register or reckon up quantity of electricity, and the latter record energy. A Board of Trade Unit (B.T.U.) is a unit of energy equal to 1,000 watt-hours. Many forms of meter register directly in Board of Trade units. Meters may be divided into self-registering meters, in which some counting mechanism records and adds up the ampere-hours or watt-hours passing through the meter, or they may require some operation of weighing or measuring to be performed before the result is known, as, for instance, in Edison's electrolytic meter.

Assuming, however, that the meter is a self-registering one, it may be tested in the following manner. Set up the meter in a proper position, either screwed to a wall or table, and connect it to a bank of incandescent lamps, so that it will record the quantity or the energy given to them. Insert an ammeter in series with the lamps, and a voltmeter across the terminals of the lamps. Start the current at a known instant, and observe the current and volts at regular and noted instants of time for a sufficiently prolonged period. The meter reading should be taken at the beginning and at the end of the run. From the instrumental readings it is possible to plot out a curve of energy or quantity given to the lamps. To do this, take a horizontal line to represent time, and mark off lengths to represent hours. At proper intervals set up perpendiculars representing the current and the watts given to the lamps at that instant, and join the tops of all these lines so as to form a current or power curve. Integrate this curve and obtain the whole area included between the time base line, the curve, and the terminal perpendiculars in terms of a unit representing one ampere-hour or one watt-hour. Compare this observed and calculated value with the meter reading. Try the same for very different currents, that is to say, do not keep the current constant, but vary it as much as possible.

If the meter is one intended only for alternating currents, then the ammeter used must be one suited for these currents, such as a Siemens dynamometer or Kelvin ampere balance, and should previously have been carefully standardized. Likewise the voltmeter must in this case be an instrument, such as a Cardew or electrostatic voltmeter, suitable for use with alternating currents. In addition to checking the accuracy of reading of the meter, several other matters should be examined. The liability of the meter to be disturbed by the presence of magnets should be tested. If the meter is of the dynamometer type, and has a shunt and series coil, the arrangement of this shunt coil should be examined to see if it is liable to become short-circuited. Also the amount of power taken upon this shunt coil should be measured by measuring the shunt coil current, and note taken whether supplier or customer pays for this power. A very small loss in this respect may mean a considerable total in the course of a year. Since there are 8,765 hours in a year, even *one watt* wasted hourly all the year round means nearly nine B.T.U., and thus 10 watts means 90 B.T.U., which at 6d. per unit amounts to a total loss of £2. 5s.

Another point which should be carefully examined is the starting power of the meter. A good 25-ampere meter should start with at least ·3 ampere. The current required to start the meter should always be less than that of one eight-candle lamp. The meter should be tried not only with a varying current, but also with constant currents from the smallest to the highest it will carry, and the meter reading compared with the observed and calculated ampere-hour or watt-hour delivery through it. If the meter contains a clock, the going-rate of this clock should be independently examined. Generally speaking, watt-hour meters are to be preferred for electric lighting work to ampere-hour meters, and meters which will act both for alternating and direct currents to meters which will act but for one kind alone.

It is impossible to indicate all the special points which the observer should investigate, as these will depend upon the construction of the meter ; but generally the points to be regarded are not merely accuracy under the laboratory test, but liability to derangement in actual every-day use over long periods. Under this head come such matters as failure of action from tarnished or dirty contacts, mercury cups or platinum points ; gradual increase of friction at bearings ; short circuits in shunt coils, or leakage of current across insulators. A satisfactory test of a meter on these points cannot be conducted in a short time ; nothing less than three months' actual use is sufficient.

Some meters are intermittent integrating meters—that is to say, the record of the current or power is not made continuously, but the meter takes a reading of the current or power every few minutes and then adds up the result. There is an objection to this class of meter wherever the current is liable to vary very rapidly, and in testing such a meter care should be taken to see that it does not stick or over read a small current after a much larger one has been passed through it.

If the motive power of the meter is an electromotor, or the meter contains electromagnets, the amount of power taken up by these should be ascertained, because, as shown above, a small but continuous absorption of power means a good deal in the course of a year.

THE CALIBRATION OF ELECTRIC METERS.

Type of meter tested –

Observation No.	Meter reading at start.	Current in amperes.	Potential difference in volts.	Time.	Meter reading at stopping.	Calculated total ampere-hours or watt-hours passed.	Meter error.

THE CALIBRATION OF ELECTRIC METERS.

Observation No.	Meter reading at start.	Current in amperes.	Potential difference in volts.	Time.	Meter reading at stopping.	Calculated total ampere-hours or watt-hours passed.	Meter error.

ELECTRICAL LABORATORY NOTES AND FORMS.

Name   *Date*

The Delineation of Hysteresis Curves of Iron.

The apparatus required is a magnetometer set up in a steady place. A long vertical magnetizing coil must be fixed near it, and means provided for regulating, measuring and reversing the current through this coil. The experiment cannot be performed in the neighbourhood of moving iron, or of iron hot-water or steam pipes, as these cause changes in the earth's magnetic force in their neighbourhood.

The Student should sketch the arrangement of the apparatus as set up.

If a rod or ring of iron is magnetized in one direction by a magnetizing force which slowly increases, and if at every stage of its increase the induction in the iron is measured, we can plot out a curve of rising induction in terms of magnetizing force. If then, after reaching a certain maximum value of the force, the direction of the force is reversed and it is diminished again and carried back past zero to an equal maximum value in the opposite direction, and then brought back again to zero, we complete what is called a *cycle of magnetization.* When the induction is plotted out in terms of the magnetizing force throughout the cycle, after going round the cycle once or twice it will be found that the induction curve is a closed loop. It is clear, therefore, that the induction lags behind the force in phase. This phenomenon is called *magnetic hysteresis,* and the curve so drawn is called the hysteresis curve of the iron. To delineate the curve we proceed as follows. On a long wooden or paper tube, two circuits of insulated wire are wound to form two magnetizing coils. This double coil should be about four feet long and one inch in internal diameter, and be wound with No. 18 S.W.G. double cotton-covered wire, and about six layers of wire be put on one coil and one on the other. This constitutes the magnetizing coil. This coil must be fixed in a vertical position against a wall running north and south. Against the same wall, at a spot six or eight inches from the coil, and at about one-quarter of its length from the end, is to be fixed a box which contains the magnetometer. This consists of a light concave mirror of silvered glass suspended by a cocoon fibre, and having three or four small magnets of watchspring fixed on the back. These magnets should not be more than a quarter of an inch in length. A lamp and scale must be provided to obtain a sharp image of an illuminated slit on a scale a metre away, and by means of the scale the angular movement of the needle can be determined. A few rods of iron are then provided, about thirty inches long and one-twentieth of an inch in diameter.

In order to prevent the current in the coil from influencing the needle directly, a compensating coil must be provided. This consists of insulated wire wound on a bobbin and placed in series with the magnetizing coil, and so situated with regard to the magnetic needle that the resultant magnetic force of the two coils at the place where the needle is is always in the direction of the undisturbed needle. When this is the case no current sent through the two coils in series can disturb the normal position of the needle.

These arrangements being made, one of the iron rods is introduced into the magnetizing coil and a small current sent through the coil. The magnetic needle then takes a deflection. The rod must be moved up or down until this deflection is a maximum. The current in the coil must then be measured by a potentiometer arrangement or by a sensitive ammeter. The experiment then consists in gradually raising the current in the coil and observing at each stage the deflection of the magnetic needle. The current is raised to a maximum value, then lowered again to zero ; next reversed and increased to an equally great negative value, and then brought back to zero. This cycle of. magnetizing force must be repeated half-a-dozen times, and the magnetizing forces and corresponding deflections of the needle observed. If the object is merely to obtain the form of the hysteresis curve, it will be sufficient to plot out these magnetizing currents as the abscissæ of a curve, and the corresponding deflections of the magnetometer as ordinates. This will give the form of the hysteresis curve plotted to an arbitrary scale.

If, however, the absolute value of the induction and force are required, we have to calibrate the magnetometer. From the known number of turns N on the magnetizing coil, and the dimensions of the same, we can calculate the magnetizing force H due to the current of A amperes in the coil from the formula

$$H = \frac{4\pi}{10} \cdot \frac{\text{ampere-turns}}{\text{length of coil}}.$$

This force is, however, added to or subtracted from the earth's vertical magnetic force which exists inside the coil.

If a long iron rod, at least 400 diameters long, is magnetised, we may consider that at a point near each end there is a magnetic pole of strength m. The whole number of lines of force coming out from the pole, or the whole induction up the centre of the rod, is equal to $4\pi m$ units. If the section of the rod is s square centimetres and the length is l centimetres, then the volume is $s\,l$ cubic centimetres. The intensity of magnetization of the rod being assumed to be uniform and equal to I, we have by the usual equation

$$I = \frac{m\,l}{l\,s},$$

where m is the strength of each pole and l is the length of the rod. Calling B the induction at the centre of the rod, we have $B\,s = 4\pi m$. Hence

$$B\,s = 4\pi\,I\,s,\ \text{or}\ B = 4\pi\,I\,;$$

hence
$$m = I\,s = \frac{B}{4\pi}\,s.$$

If the upper magnetic pole of the rod is on a level with the magnetometer needle, and if O P is the distance from the needle to the pole, and likewise if O P¹ is the distance from the needle to the lower magetic pole of the rod, it is easily seen that the magnetic force due to the two poles of the rod are equal to a force F acting on the magnetic needle, where

$$F = \frac{m}{(O\,P)^2} - \frac{m}{(O\,P^1)^2}\,\frac{O\,P}{O\,P^1}.$$

If this force, acting on the suspended needle, causes it to make a deflection θ when F acts at right angles to the earth's horizontal magnetic force H, we have $F = H \tan \theta$.

Hence, since
$$m = \frac{B}{4\pi} s,$$

we have
$$H \tan \theta = \frac{B}{4\pi} s \left(\frac{1}{(O P)^2} - \frac{O P}{(O P^1)^3} \right),$$

and B is determined in terms of quantities which can all be measured easily, knowing the value of the earth's horizontal force. If the rod is at least 400 diameters long, the influence of the poles induced in the rod in weakening the effective magnetizing force will be negligible, but this is not the case if it is a short rod.

If, then, the earth's vertical magnetic force V is known, the total magnetizing force M, acting on the long rod, is equal to

$$\frac{4\pi}{10} \frac{\text{ampere-turns}}{\text{length of coil}} \pm V,$$

where the term " length of coil " means the length between the cheeks of the magnetizing spiral. Hence we can calculate and find the absolute values of M and B, and plot the hysteresis curve in absolute units.

This experiment can only be well carried out in a room where H and V are both constant and accurately known.

Provided, however, these conditions can be fulfilled, we can determine in absolute value the area of the hysteresis loop for the iron used.

The object of putting the two coils on the bobbin is to be able to neutralize the earth's vertical force by sending a suitable constant current through one of the coils. The earth's vertical force is known to be neutralized when the iron is left perfectly non-magnetic, by causing an alternating current, gradually decreased in strength to zero, to flow through the other coil, and when, after this process, a slight displacement of the rod does not affect the magnetometer needle.

For full details of the magnetometer method the Student should consult Chapter II. of the text-book on " Magnetic Induction in Iron and other Metals," by Prof. J. A. Ewing (" Electrician " Series).

THE DELINEATION OF HYSTERESIS CURVES OF IRON.

Length of iron rod used ⁼

Diameter of iron rod used ⁼

Distance of magnetometer needle from scale = d =

No. of turns of wire in magnetizing coil = N ⁼

Length of magnetizing coil = L ⁼

Earth's horizontal force = H ⁼

Earth's vertical force = V ⁼

Observation No.	Excursion of spot of light of magnetometer = x.	Tangent of angle of deflection of needle = $\tan \theta$ $\frac{x}{d} = \tan 2\theta$.	Current in amperes flowing in magnetizing coil = A.	Value of $\frac{4\pi}{10} \frac{A N}{L}$ = magnetizing force due to current.	Value of H $\tan \theta$.	Value of V.	Calculated value of B.

ELECTRICAL LABORATORY NOTES AND FORMS.

Name.   *Date*

The Examination of a Sample of Iron for Hysteresis Loss.

The apparatus required for these experiments is a sensitive electrostatic voltmeter adapted for alternating current measurements. A steady source of alternating currents of various frequencies must be available. Some thin sheet-iron must be made up into bundles, each sheet being separated from the next by thin paper. A Siemens dynamometer is also required. Four magnetizing coils, each wound with two circuits, are required for magnetizing the samples of iron.

The Student should carefully sketch the arrangement of the apparatus.

If a sample of sheet-iron, as used for transformers, is submitted for test, the simplest method by which to examine it is to make up a choking core with a known weight of the iron. Cut a large number of strips of iron, say one or two inches in width and twelve or eighteen inches long. Divide these strips into four bundles and weigh the iron. Proceed then to paper each strip of iron on one side with a strip of tissue paper pasted on, and make up these strips into bundles, fastening the strips together by binding them with tape, but leaving a couple of inches at each end free. Four magnetising coils have then to be wound by winding covered wire on a paper tube with wooden cheeks. Two well-insulated wires should be wound together on each tube. The number of turns on each coil must be known. These coils may be wound with two No. 16 double-cotton covered wires laid on together, and must have, at least, 20 turns per centimetre of length.

Each coil is then slipped on to a bundle of strips, and the coils arranged in a square form. The ends of the strips of iron where the bundles meet at the corners must then be interlaid or sandwiched in between each other, and squeezed together with a clamp so as to make a good magnetic joint. We have then a square iron frame formed of laminated iron of known mass, and wound over with two circuits each having a known number of turns of wire. Measure the length of the magnetic circuit by taking the mean length of the square iron core, and calculate the wire turns per unit of length of the magnetic circuit. Calculate also from the known thickness and width of the iron the cross-sectional area of the iron core at the centre of each bundle. Join in series in the same direction the separate circuits on each bobbin so as to make one complete primary and secondary coil, and connect the primary coil with a source of supply of alternating current through a variable resistance. Insert a standardized

Siemens dynamometer in circuit with this primary coil, and arrange the electrostatic voltmeter so that it can be connected quickly either to the terminals of the primary or the secondary circuit. Measure the electrical resistance of the primary circuit, and ascertain if there is any leakage between the two circuits. If so, the double-wound coils must be re-insulated. The frequency of the alternating current employed must be known. These arrangements being made, we proceed to measure with the dynamometer the primary current, and with the electrostatic voltmeter the potential difference at the terminals of the primary and secondary coils. These readings of course give us the mean square ($\sqrt{\text{mean}^2}$) value of these quantities. They are connected as follows with the hysteresis loss in the iron core :—

Let e_1 be the value *at any instant* of the primary potential difference (P.P.D.), and e_2 that of the secondary circuit, and let N_1 and N_2 be the number of windings on these circuits. Let i be the instantaneous value of the primary current, and b that of the induction in the core. These quantities are related as follows :—

$$N_1 S \frac{d b}{d t} + r i = e_1 \quad . \quad . \quad . \quad . \quad . \quad . \quad . \quad (i.)$$

$$N_2 S \frac{d b}{d t} = e_2, \quad . \quad (ii.)$$

where S is the cross sectional area of the core.

From the equations $(i.)$ and $(ii.)$ eliminate S and $\frac{d b}{d t}$, and we obtain the equation

$$\frac{N_1}{N_2} e_2 = e_1 - r i. \quad . \quad . \quad . \quad . \quad . \quad . \quad . \quad (iii.)$$

Square this equation and multiply all through by $d t$, and we get

$$\left(\frac{N_1}{N_2}\right)^2 e_2^2 = e_1^2 + r^2 i^2 - 2 r e_1 i,$$

or

$$e i d t = \frac{1}{2} r i^2 d t + \frac{1}{2 r} e_1^2 d t - \left(\frac{N_1}{N_2}\right)^2 e_2^2 d t.$$

The expression on the left-hand of this last equation represents the instantaneous value of the whole power given to the coils and core, and if we write it

$$e i d t - r i^2 d t = \frac{1}{2 r} e_1^2 d t - \frac{1}{2 r} \left(\frac{N_1}{N_2}\right)^2 e_2^2 d t - \frac{1}{2} r i^2 d t,$$

we see that the expression on the left hand now represents the loss in the iron core alone. If, instead of taking instantaneous values, we take mean values throughout a complete period, and write E_1^2, E_2^2, I^2 for the mean values of e_1, e_2, and i respectively, and H for the mean loss in the core, we obtain the equation

$$H = \frac{1}{2 r} E_1^2 - \frac{1}{2 r} \left(\frac{N_1}{N_2}\right)^2 E_2^2 - \frac{1}{2} r I^2.$$

The quantities E_1^2, E_2^2, I^2 are all given by the instruments. If N_1 and N_2 are equal, the equation reduces to

$$H = \frac{1}{2 r} \{ E_1^2 - E_2^2 \} - \frac{1}{2} r I^2. \quad . \quad . \quad . \quad . \quad (iv.)$$

The loss in the core is therefore known when we know the terminal volts (mean square value) of the primary and secondary circuits, and the value of the primary current and primary circuit resistance.

We require to know, however, the value of the mean induction in the core.

If the supply of current is being furnished by an alternator having an electromotive force curve, not far from a simple sine curve, then it becomes a very simple matter to calculate the value of the induction. For if the primary electromotive force curve is not very different from a simple sine curve, then the curve of induction will be still more nearly a simple sine curve. Let B be the maximum value of the induction, and let $p = 2\pi n$, where n is the frequency, then

$$b = B \sin p\, t,$$

hence

$$\frac{d\,b}{d\,t} = p\,B \cos p\, t, \quad \cdot \quad \cdot \quad \cdot \quad \cdot \quad \cdot \quad (v.)$$

and from equation (ii.) $\qquad e_2 = N_2\, S\, p\, B \cos p\, t.$

Hence the maximum value of the secondary electromotive force E_2^1 is equal to $N_2\, S\, p\, B$,

or

$$B = \frac{E_2^1}{N_2 S\, p}. \quad \cdot \quad \cdot \quad \cdot \quad \cdot \quad \cdot \quad \cdot \quad \cdot \quad \cdot \quad (vi.)$$

If B^1 is the $\sqrt{\overline{\text{mean}^2}}$ value of the induction density, we can calculate the value of B^1, when we know E_2, N_2, S and p, from the equation $B^1 = \dfrac{E_2}{N\, S\, p}$.

The value of the primary current should be so adjusted that $B^1 = 3,000$ C.G.S. units per square centimeter, this being a very usual value for B^1 in the cores of transformers.

The primary pressure should therefore be varied, and a number of readings taken of the value of E_1, E_2, and I, and the loss H in core, and the value of the induction in the core calculated for each set by equations (iv.) and (vi.). These values of H and B should be set out in a curve. From the known weight and volume of the core the hysteresis loss per pound or per kilogramme of the iron can be calculated. In good iron it should not exceed 0·5 watt per pound. It will be seen that, since in equation (iv.) we have to take the difference of the squares of two quantities, these quantities themselves must be measured with great accuracy. It is essential, therefore, to have a very steady alternating current, and to take the readings with great care ; otherwise a very small percentage error in the measurement of E_1 and E_2 will make a very large percentage error in the value of $(E_1^2 - E_2^2)$.

On the question of hysteresis loss in iron the Student may further consult "The Alternate Current Transformer" (Fleming), Vol. II., page 479 ("Electrician" Series).

THE MEASUREMENT OF HYSTERESIS LOSS IN IRON.

Thickness of iron strips =
Total weight of iron =
Section of iron core = S =
Number of turns on primary coil = N_1 =
Number of turns on secondary coil = N_2 =
Frequency of the current = n
Resistance of primary coil = r
Value of $2\pi n = p$ =

Observation No.	Primary P.D. $\sqrt{\text{mean}^2}$ value = E_1.	Secondary P.D. $\sqrt{\text{mean}^2}$ value = E_2.	Primary current $\sqrt{\text{mean}^2}$ value = I.	Induction $\sqrt{\text{mean}^2}$ value B^1.	Total loss in iron core in watts = H.	Hysteresis loss per pound in core in watts.

ELECTRICAL LABORATORY NOTES AND FORMS.

Name   *Date*

The Determination of the Capacity of a Concentric Cable.

The apparatus required for these tests is a length of at least half-a-mile of concentric cable, a standard condenser, and a ballistic galvanometer; discharge keys and an insulated battery are also necessary.

The Student is recommended to sketch carefully the arrangement of the circuits and apparatus.

Every electric conductor has three qualities, viz., its electrical resistance, depending on the material and size of the conductor ; its electrical inductance, depending on the form and shape of the conductor and the magnetic permeability of the surrounding conducting material ; and its electrical capacity, depending on its form and surface, and the specific inductive capacity of the surrounding dielectric, or non-conductor. If an insulated cable with a single stranded conductor is coiled up and placed in a tank of water it has a certain capacity, but if uncoiled and laid out straight on the ground its capacity would have a different value.   In general there is not much practical information to be gained by measuring the capacity of a length of single cable, but in the case of a concentric cable, which consists of one stranded copper conductor wholly surrounding another, but separated by a definite thickness of insulation, there is a fixed and constant capacity per unit of length which is independent of the way in which the cable is coiled or laid.   Concentric cables are almost exclusively used for the conveyance of alternating currents, because, since one conductor wholly incloses the other, there is no external magnetic field, and, therefore, no possibility of disturbance of neighbouring telephone or telegraph lines by induction.   On the other hand, concentric cables present certain peculiarities of their own when used for the conveyance of alternating currents.   There is a current which flows across the dielectric, or non-conductor, which is called the *capacity current* of the cable.   This current is not in step with the impressed electromotive force, but in advance of it its phase and its magnitude can be calculated from the formula

$$I = \frac{C\,V\,p}{10^8},$$

where I is the maximum or mean square ($\sqrt{\text{mean}^2}$) value of this capacity current, and V is the same function of the difference of potential between the inner and outer members of the cable, C is the capacity of the cable in microfarads, and $p = 2\pi n$, where n is the frequency of the current.

It is always desirable to know what capacity current we may expect to obtain when a given length of some concentric cable is laid down to take an alternating current of known frequency and at a known pressure, because this current has to flow into the cable at the dynamo end although nothing comes out at the other, and this current is always present, although not in step with the conductive current flowing along the cable. Thus, for example, if a concentric cable has a length of two miles and a capacity of ·3 of a microfarad (m.f.d.) per mile, and is worked at 2,000 volts, at a frequency ($\sim$) of 100, we have for the mean-square value of the capacity current the number—

$$I = \frac{2 \times \cdot 3 \times 2000 \times 2 \times 3\cdot 1415 \times 100}{10^6} = \cdot 75 \text{ ampere :}$$

in other words, current enough to light a 20-c.p. incandescent lamp would flow into such a cable at the alternator end, although the far end was completely insulated and open-circuited. This must always be remembered, otherwise it may be mistaken for bad insulation.

The above equation, $I = \frac{C V p}{10^6}$, affords a method of measuring C when I, V, and p are known, but if the working pressure V is not available, then the capacity of the cable can be obtained by comparing it with that of a standard condenser. To do this, provide a ballistic galvanometer and take carefully its logarithmic decrement (*see* Elementary Form No. 5). Then join up a battery, the standard condenser, a galvanometer and discharge key, as shown in the diagram in Fig. 1. When the key K_1 is raised, the condenser is charged at the battery, and when the key is depressed the condenser is discharged through the galvanometer. On depressing the key K_1, the needle of the galvanometer receives an impulse, which causes it to make a "throw." A little experience will, however, show the Student that with most condensers the throw varies with the time of charging and the interval between charge and discharge, and that the capacity of the condenser is not a sharply-defined quantity. A first approximation, however, to the capacity of the cable may be obtained by charging successively first the standard condenser S and then the cable C at the same potential, and then discharging them instantly through the ballistic galvanometer. The capacities of the cable and condenser are then in the ratio of the sines of half the angles of throw of the needle. In order to get fairly good results, however, the capacities of the cable and condenser must be nearly the same, the galvanometer must not be shunted, and the "throws" must be about equal in each case, and of such a magnitude that they can be read to one per cent.

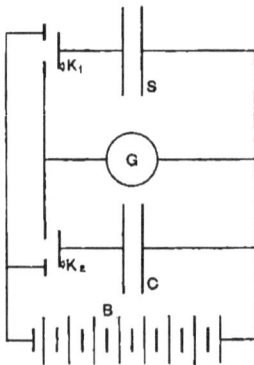

Fig. 1.

There are, however, considerable difficulties in getting close results in this way, and a better method to adopt is the following: It is known as the method of mixtures. A battery of 50 cells, well insulated, which may preferably be small lithanode secondary cells (*see* Fig. 2), has its poles closed by a very high resistance $a b$, say 100,000 ohms. The standard condenser S and cable C are joined up, as shown in Fig. 2, and keys provided, by means of which the following operations can be performed :—

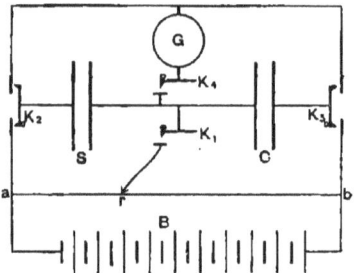

Fig. 2.

When the keys K_1 K_2 K_3 are put down, they join the outside of the condenser S to one pole of the battery, and the outer conductor of the cable C to the other pole of the battery, and the inside conductors of both condenser and cable are joined through to some point, r, on the slide wire joining the poles of the battery. On raising the keys it breaks these connections, and puts together the outside conductors of the cable and condenser, and immediately afterwards, on depressing the key K_4, we can connect them to the terminal of the galvanometer as shown. The experiment consists in varying the magnitudes of the two sections $a r$ and $r b$ of the resistances $a b$, until the galvanometer indicates no deflection when the above key operation is performed. If C is the capacity of the condenser and C^1 that of the cable, and if V and V^1 are the potentials at which they are charged, the quantities put into the condensers are respectively C V and C^1 V^1, but $V : V^1 :: R : R^1$, where R and R^1 are the ratio of the resistances of the two sections into which the whole resistance is divided. Hence, if R and R^1 are varied until C V = C^1 V^1, we have then equal quantities of charge in the two condensers, and on putting them together these charges neutralize one another and nothing remains to affect the galvanometer.

Hence, under these conditions, $C V = C^1 V^1$, or

$$\frac{C}{C^1} = \frac{V^1}{V};$$

but

$$\frac{V^1}{V} = \frac{R^1}{R},$$

therefore

$$\frac{C^1}{C} = \frac{R}{R^1},$$

or

$$C^1 = C \frac{R}{R^1},$$

and C^1 is determined in terms of the capacity C and the resistances R and R^1. This being a mill method is preferable to the other.

The Student should experiment in this manner with a length of concentric cable, and determine in the form of curves the relation between the capacity and the time of charging. Also he will note that, after the cable has been once discharged, a second discharge can be obtained after a time. This is called a residual discharge, and, with certain insulating materials, a long series of successive residual discharges can be obtained.

If the conductors are cylindrical and concentric, and if the diameter of the inside of the outer conductor is D centimetres, and the diameter of the outside of the inner conductor is d centimetres, the capacity in microfarads of the cable can be calculated from the formula

$$C \text{ (in m.f.d.)} = \frac{K l}{9 \times 10^5 \times 4 \cdot 605 \times \log_{10} \frac{D}{d}}$$

where K is the specific inductive capacity of the material and l is the length in centimetres of the cable. If $l =$ one statute mile = 160,933 centimetres,

$$C \text{ (in m.f.d.)} = \frac{K}{26 \log_{10} \frac{D}{d}}$$

For paper insulation K has a value about 2·9, and for india-rubber a value about 4.

THE CAPACITY OF A CONCENTRIC CABLE.

Type of cable used =
Length =
Inner conductor =
Outer conductor =
Insulation =
Maker's name =

Observation No.	Volts at which charge is made.	Capacity of standard condenser used.	Throw on ballistic galvanometer given by charge of cable.	Throw on ballistic galvanometer given by charge of standard condenser.	Time of charge of cable.	Capacity of cable.

No. 40.—ADVANCED (No. 20).

Name   *Date*........................

The Hopkinson Test of a Pair of Dynamos.

The plant required for this test is a pair of similar continuous-current dynamos, preferably shunt machines. These must be bolted down on a bedplate with shafts coupled in one line. Resistances must be provided in the fields of each machine. A secondary battery must be available for supplying current to make up the difference of power. Connecting leads must be taken from the machine which acts as a dynamo to that which acts as a motor. Three voltmeters and three ammeters will be required, or else a potentiometer set.

The Student is recommended to sketch carefully the arrangement of the circuits.

If a dynamo machine is being tested for efficiency in the ordinary manner, by measuring the power applied to it by a transmission dynamometer and the output by electrical instruments, the error made in determining the efficiency is proportional to the error made in measuring the power supplied. If, however, two identical machines are available, it is possible to apply Dr. Hopkinson's method, by which a greater degree of accuracy may be obtained. The simplest method of doing this is as follows : Let the two identical dynamo machines be bolted down on the same bedplate, and let them have their shafts in one line and coupled together. Let the one be called the dynamo (D), and the other the motor (M). Take leads from the terminals of D and join them to the terminals of M, and insert in the circuit of one of these leads a secondary battery so joined in that its electromotive force aids the electromotive force of the dynamo in rotating the motor in the proper direction to self-excite the coupled dynamo Let the field-magnets of D and the field-magnets of M be excited from the brushes of the dynamo and motor respectively. Insert ammeters in the circuit of the armatures of the motor and dynamo. Call the armature current flowing between the machines A_1. Insert ammeters in the fields of both D and M, and call these currents a_1 and a_2 respectively. Insert voltmeters across the terminals of each machine. Call the volts across the brushes of the dynamo V_1, and that across the terminals of the motor V_2, and provide means for regulating the speed of the machines and their excitation by inserting resistance in the circuit of the field-magnets of both machines.

The test is started by varying the number of cells in the secondary battery until the current which passes through the armature circuit, viz. A_1, is that corresponding to the full load of the dynamo. The resistances in the fields of the machines are then varied until the machines acquire their normal speed, and the instruments are then read. The arrangement of the circuits may be understood from the annexed diagram. Since the current which comes out of the dynamo is A_1 and the terminal P.D. is V_1, the power coming out of the dynamo is $A_1 V_1$ watts. If V_3 is the P.D. down the secondary battery, then $A_1 V_3$ is the power supplied by the battery. Since V_2 is the P.D. at terminals of the motor, $A_1 V_2$ is the power supplied to the motor.

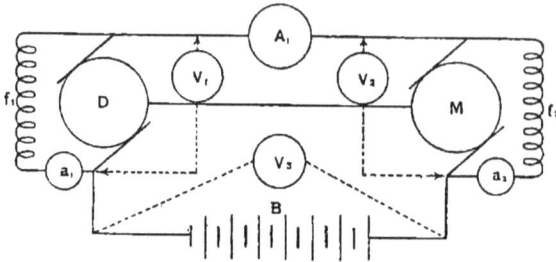

Fig. 1.

Then we have the following measured powers. The power delivered from the dynamo is $A_1 V_1$ watts. The power delivered to the motor is $A_1 V_2$ watts. The difference between these powers, viz., $A_1 V_2 - A_1 V_1$, must be supplied by the secondary battery, and must be equal to $A_1 V_3$. Hence

$$A_1 V_3 = A_1 V_2 - A_1 V_1,$$

or

$$V_3 = V_2 - V_1.$$

The power $A_1 V_3 = A_1 (V_2 - V_1)$ is required to supply the power lost in both machines taken together. If we assume that the total losses in both machines are equal, we may arrive at the figure representing the efficiency of each machine as follows :—

The efficiency e_1 of the dynamo D is the ratio of the power coming out of it to that going into it.

The power coming out of D is $A_1 V_1$ watts into the external circuit and $a_1 V_1$ watts into its own field-circuits.

The total power given out by the dynamo on the external circuit is, therefore, $A_1 V_1$ watts.

The power put into the dynamo is, therefore, $\dfrac{A_1 V_3}{2} + A_1 V_1 + a_1 V_1$ watts.

The dynamo efficiency is, therefore,

$$e_1 = \frac{A_1 V_1}{\dfrac{A_1 V_2}{2} + \dfrac{A_1 V_1}{2} + a_1 V_1} \qquad . \qquad (1)$$

The power put into the motor is $A_1 V_2 + a_2 V_2$ watts.

The power given out by the motor is, therefore, equal to $A_1 V_2 + a_2 V_2 - \tfrac{1}{2} A_1 V_3$,

or is equal to

$$\frac{A_1 V_2 + A_1 V_1}{2} + a_2 V_2.$$

Hence the efficiency of the motor is

$$e_2 = \frac{\tfrac{1}{2} A_1 V_2 + \tfrac{1}{2} A_1 V_1 + a_2 V_2}{A_1 V_2 + a_2 V_2}. \qquad (2)$$

If for the moment we neglect the exciting currents a_1 and a_2 in comparison with A_1, we see that the efficiency of the dynamo e_1 becomes

$$e_1 = \frac{2 V_1}{V_2 + V_1} \qquad \ldots \ldots \quad (3)$$

and the efficiency e_2 of the motor is $e_2 = \frac{V_2 + V_1}{2 V_2}$,

and that

$$e_1 e_2 = \frac{V_1}{V_2}. \qquad \ldots \ldots \ldots \quad (4)$$

If, instead of assuming that the power supplied by the secondary battery is equally divided between the two machines, we assume that the efficiency of the machine used as a dynamo is equal to the efficiency of the machine used as a motor and call this equal efficiency e, and if we call the power supplied on the shaft common to the two machines P, then we have for the dynamos—

$$e = \frac{A V_1}{P},$$

neglecting the power required to excite ; and for the motor—

$$e = \frac{P}{A V_2}.$$

Hence

$$e = \sqrt{\frac{V_1}{V_2}}. \qquad \ldots \ldots \ldots \quad (5)$$

Comparing equations (4) and (5), we see that $\sqrt{e_1 e_2} = e$. In other words, the efficiency found in the assumption that the machine which functions as a dynamo has the same efficiency as the machine which functions as a motor is the geometrical mean of the several efficiences found by assuming that the *losses* in the two machines are equal when running thus coupled. To obtain a definite result we must make one of these two suppositions.

The above method is the most accurate method of finding the efficiency of a dynamo and motor when they are sufficiently alike to make one of the two above suppositions nearly true ; and it has this advantage : that all the measurements to be made are electrical measurements, and hence can be made very accurately.

If a Crompton potentiometer is available, the above readings of currents and volts may be made very accurately by inserting in the field and armature circuits low resistances of such value that, with the normal currents through them, the fall of potential down these resistances is not more than 1·5 volts. Then from the ends of these resistances potential wires are brought to the potentiometer, and the three currents, A_1, a_1, and a_2, and the two voltages, V_1 and V_2, very quickly read in succession. The voltages must, of course, be read with the aid of a divided resistance.

The Student should in this way make a test of a pair of coupled machines, obtaining the efficiency of each at various loads, and plotting a curve showing the efficiency as a function of the current through the armature.

HOPKINSON TEST OF A PAIR OF COUPLED DYNAMOS.

Dynamo No.   by
Motor   No.   by
Number of cells of secondary battery used =

TABLE I.

Observation No.	Speed revolutions per minute.	Armature current A_1.	Dynamo volts V_1.	Motor volts V_2.	Dynamo field-current a_1.	Motor field-current a_2.	Battery volts V_3.

TABLE II.

Observation No.	Watts put into dynamo W_1.	Watts taken out of dynamo W_2.	Efficiency of dynamo e_1.	Watts put into motor W_3.	Watts taken out of motor W_4.	Efficiency of motor e_2.	Current A_1.

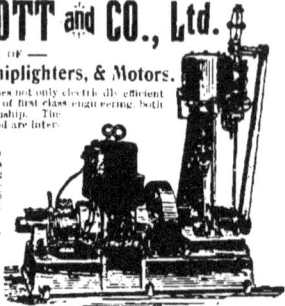

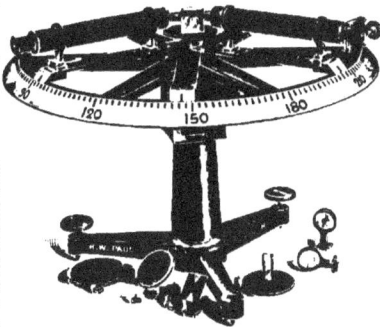

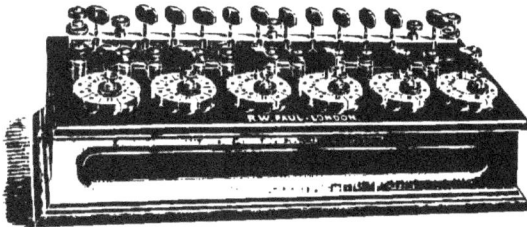

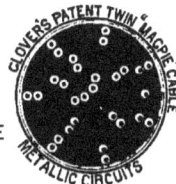

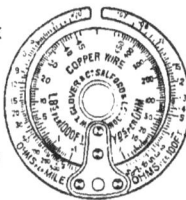

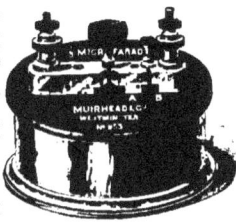

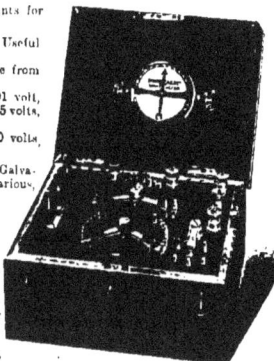

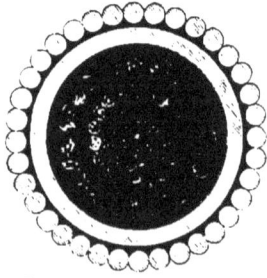

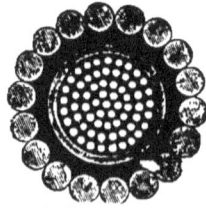

www.ingramcontent.com/pod-product-compliance
Lightning Source LLC
Chambersburg PA
CBHW021523210326

41599CB00012B/1363